新手入廚

鄭慧芳 編著

萬里機構・飲食天地出版社出版

Chinese Cooking for Beginners

學中菜

前言

現代都市女性生活繁忙，工作之餘又要照顧家庭，但憑着一份愛心，為了使家人食得健康又開心，不少人曾到烹飪學校學習烹飪。近幾年，一股 DIY 自製糕餅熱潮襲港，引起一些年青人對烹飪的興趣。從前，他們以為煮中菜是媽媽的專利，近年我在煤氣烹飪中心教過不少年青白領人士，不但有女性，還有很多是男兒呢！雖然他們很多是廚藝新手，經過指導後，身手也不錯。

我常説："只要學會了基本功，手執一本烹飪書也可給你技術支援，做出大受歡迎的菜式。"今次我這本《新手入廚學中菜》是針對一些入廚新丁，輯錄了五十多款菜式，雖然是製作簡易的家常小菜，我也會詳細描述製作過程，還會提出可能犯的錯誤。若倍加注意及跟隨步驟，便能得心應手，一定做得稱心滿意(無論如何，自己煮的菜式是最新鮮、最美味的)。這本書包括有：蛋菜豆腐、鮮魚海產、禽畜肉類等幾個類別，務求做到菜式多元化。

《幸福的甜品》是我的第一本作品，深得學生及讀者歡迎，從而激發我出版另一本中菜食譜，在各方的熱烈支持下，加上家人及朋友的鼓勵，尤其是得好朋友 Mindy 胡玉玲小姐的幫忙，製作得以順利完成。謹此致謝。

鄭慧芳

FOREWORD

Metropolitan ladies lead a busy life. Other than working, they have to take care of their families, ensuring good health and happiness of their loved ones. In this light, some of them learn cooking skills at cookery schools so as to please their families with heart-warming food. In recent years, with the popularity of DIY cakes, some young people have become interested in cooking. Instead of relying on their mothers to cook for them, a number of young executives, both females and males, come to learn from me at the cookery centre. Although many of them are novices, they have all mastered the basic skills after some training.

I always say, "It only takes some elementary skills and a cookbook for technical support to make a number of popular dishes." Therefore, this book is specially dedicated to the beginners of cooking with recipes of more than 50 simple homemade dishes. Although the dishes are easy to make, I have listed out the procedures in details and have pointed out the common mistakes made by many people. If you follow the steps stated carefully, I am sure you can successfully make the dishes you like. This book contains a variety of dishes including recipes for dishes of vegetables and tofu, fresh seafood, and poultry and livestock.

The success of my debut cookbook **Chinese Desserts**, which is well-received by both my students and readers alike, prompted me to write another cookbook on Chinese dishes. With the enthusiastic support from all as well as the encouragement from my family and friends, in particular my good friend Miss Mindy Wu, who gave me a helping hand, I have managed to complete this book successfully. Thank you.

Lilian Cheng

目錄 | CONTENTS

46 76 104 112

三文魚金華豆腐

Steamed Tofu with
Salmon and Ham

預備時間：**10分鐘** ■烹調時間：**5分鐘** ■份量：**4人**

■ Preparation time：**10 mins**

■ Cooking time：**5 mins**

■ Serves：**4**

材料：
鮮三文魚柳4兩（150克）
金華火腿1/4兩（10克）
蒸煮滑豆腐1件
蔥1條

調味：
鹽1/8茶匙
油1茶匙

豉油汁：
水1湯匙
生抽2湯匙
糖1茶匙
老抽1/4茶匙
麻油1/2茶匙

Ingredients:
150g fresh salmon fillet
10g Jinhua ham (Chinese cured ham)
1 pc soft tofu for steaming
1 sprig spring onion
Seasonings:
1/8 tsp salt
1 tsp oil
Sauce Ingredients:
1 tbsp water
2 tbsp soy sauce
1 tsp sugar
1/4 tsp dark soy sauce
1/2 tsp sesame oil

↘TIPS 貼士:

盒裝豆腐可即食，所以此菜只需要計算蒸魚時間便可（三文魚過熟又不美味，只需略蒸便可）。如果想價錢經濟實惠，可在壽司店買三文魚的魚肉邊，再將魚肉切粒，鋪在豆腐面上，再灑上金華火腿蒸，營養豐富，小朋友適宜常吃。

As tofu in packet is ready to eat, it just takes time to steam the fish (steam the salmon briefly otherwise it will become less tasty). If you want to have quality salmon at a bargain, you may buy the fillet trimmings from sushi shops, cut it into dices, spread over the tofu, and then sprinkle some Jinhua ham on top before steaming. It is such a nutritive dish that is ideal for children.

做法：

1. 蔥切粒；金華火腿切幼長條。

2. 三文魚斜切大片，塗上調味料，加適量金華火腿條，捲上，備用。

3. 豆腐修切成八件方件，吸乾水，排碟上，豆腐面塗上少許鹽及生粉，把三文魚卷放豆腐面，隔水以中火蒸3分鐘至熱透。

4. 燒1湯匙油，下蔥粒及豉油汁煮滾　淋豆腐邊，趁熱品嚐。

Method:

1. Cut spring onion into dices; cut Jinhua ham into strips.

2. Cut salmon fillet diagonally into large pieces, smear with seasonings, put on a few ham strips, roll it up and then set aside.

3. Cut tofu into 8 cubes, pat dry, line on a dish, smear a little salt and cornstarch on the surface, put a salmon roll on top of each tofu cube, and then steam above hot water on medium heat for 3 mins until cooked.

4. Heat 1 tbsp of oil, add in spring onion dices and sauce ingredients until boils. Pour it along the sides of the tofu. Serve hot.

叉燒涼瓜炒蛋

Scrambled Egg
with BBQ Pork and
Bitter Melon

預備時間：**15分鐘** ■烹調時間：**10分鐘** ■份量：**4人**

■ Preparation time：**15 mins**
■ Cooking time：**10 mlns**
■ Serves：**4**

材料：
涼瓜6兩 (240克)
叉燒2兩 (80克)
雞蛋5隻
葱1條

調味：
鹽½茶匙
糖1茶匙
雞粉¼茶匙
胡椒粉及麻油少許
油½湯匙

Ingredients:
240g bitter melon
80g BBQ pork
5 eggs
1 sprig spring onion

Seasonings:
½ tsp salt
1 tsp sugar
¼ tsp chicken powder
A little pepper and sesame oil
½ tbsp oil

⊠TIPS 貼士：

將苦瓜切成薄片後下鹽醃過，拖水片刻，苦味便會減低。與雞蛋配合同煮的話，也可中和苦瓜的苦味，連小朋友也會接受。若喜愛吃苦瓜的甘苦味的話，可先將苦瓜切片即落油鑊炒透，再放回蛋內同煎即可。

To remove the bitterness of bitter melon, simply cut it into thin slices, marinate with salt and then blanch in boiling water briefly. You can also cook with eggs to neutralize the bitter taste so that it will become acceptable to children as well. If you like the strong bitter taste, you may cut the melon into thin slices and then stir-fry with oil thoroughly before steaming it with eggs.

做法：

1. 葱切粒；叉燒切條。
2. 涼瓜開半，挖去瓜瓤，切薄片，加1茶匙鹽拌勻，同醃10分鐘，放半鑊滾水內拖水(即飛水)1分鐘，取出，沖凍水，搾乾水，加1湯匙生粉拌勻。
3. 雞蛋打散，加調味料、葱粒、叉燒及涼瓜片拌勻。
4. 燒熱鑊加3湯匙油，慢慢倒入蛋漿料，用筷子拌炒至凝固及熟透，上碟。

Method:

1. Cut spring onion into dices; cut BBQ pork into strips.
2. Cut bitter melon in half, scrape the core and seeds, cut into thin slices, add in 1 tsp of salt, mix well and marinate for 10 mins. Blanch it into a pot of half-filled with boiling water for 1 min, take it out, rinse with cold water, squeeze out excess water, add in 1 tbsp of cornstarch and mix well.
3. Beat the eggs thoroughly, add in seasonings, spring onion dices, BBQ pork and bitter melon, and mix well.
4. Heat up a wok, add in 3 tbsp of oil, pour in egg batter gradually, stir with chopsticks until solidified and cooked. Transfer the fried eggs to a dish.

日式海鮮蒸碗蛋

Japanese Style Steamed Egg with **Seafood**

預備時間：**10分鐘** 烹調時間：**10分鐘** 份量：**4人**
- Preparation time：**10 mins**
- Cooking time：**10 mins**
- Serves：**4**

材料：
蟹味菇½盒
日本蟹柳3條
鳴門卷8片
鮮蝦肉8隻
雞蛋3隻
滾水約270毫升

調味：
鰹魚粉1½茶匙
鹽¼茶匙
糖⅛茶匙
胡椒粉少許

Ingredients:
½ packet crab mushroom
3 Japanese crab meat sticks
8 slices Naruto fish rolls
8 fresh shrimps
3 eggs
270ml boiling water

Seasonings:
1½ tsp Bonito powder
¼ tsp salt
⅛ tsp sugar
A pinch of pepper

TIPS 貼士：

蒸茶碗蛋的好處是每人一份，內裏餡料可隨意搭配。鰹魚粉是日式風味的關鍵，一般在日式超市買得到，因每套碗具不同，蒸的時間可在慢火期間加長，甚至可熄火後焗至熟透，切忌用大火蒸過時。

The beauty of steaming eggs in bowls is that it can be served individually to each guest. Besides, you can also mix and match the fillings to your own taste. Bonito powder, the key to its Japanese flavour, is widely available in Japanese supermarkets. Since bowl sets are different from one another, you may have to adjust the steaming time required. You may even leave the bowls covered with lids after turning off the heat to cook but don't over cook it on high heat.

做法：

1. 滾水加調味料拌勻，待涼。

2. 雞蛋拂勻，與以上調味料拌勻，用篩濾過，備用。

3. 蟹柳開半拆絲；鳴門卷切薄片，蟹味菇剪去菇腳，沖水；蝦肉挑去黑腸，沖洗乾淨。

4. 燒半煲滾水，分別把蟹柳、鳴門卷、蝦肉及蟹味菇略拖水，瀝乾水，放四個小碗內，注入蛋液，隔水以中慢火先蒸4分鐘，再轉用慢火蒸3分鐘及至熟，便成。

Method:

1. Mix seasonings with boiling water, set aside to cool.

2. Beat the eggs thoroughly, mix with the seasonings, filter with a sieve and then set aside.

3. Shred crab meat sticks; cut Naruto fish roll into thin slices; cut the stems of the mushrooms and then rinse with water; remove the intestines of the shrimps and rinse with water thoroughly.

4. Bring half pot of water to a boil, blanch crab meat, Naruto fish roll, shrimp meat and crab meat mushrooms in it separately, strain, put them into four different small bowls, and then pour in egg mixture. Steam above water on medium to low heat for 4 mins, and then steam on low heat for another 3 mins until cooked. Ready to serve.

家常豆腐

Braised Tofu with Pork
and **Mushrooms**

預備時間：**15分鐘** ■烹調時間：**15分鐘** ■份量：**4人**

■ Preparation time：**15 mins**

■ Cooking time：**15 mins**

■ Serves：**4**

材料：
布包豆腐2件
枚頭瘦肉2½兩（100克）
草菇3兩（120克）
青豆2湯匙
薑米1茶匙
蒜蓉1茶匙
蔥1條

醃料：
鹽及糖¼茶匙
生抽½茶匙
生粉1茶匙
水1湯匙
胡椒粉、麻油少許

芡汁：
水½杯
糖1茶匙
生粉2茶匙
生抽1茶匙
蠔油1湯匙

Ingredients:
2 pcs wrapped soft tofu
100g lean pork
120g straw mushrooms
2 tbsp green peas
1 tsp minced ginger
1 tsp minced garlic
1 sprig spring onion

Marinade Ingredients:
¼ tsp salt and sugar
½ tsp soy sauce
1 tsp cornstarch
1 tbsp water
A little pepper and sesame oil

Thickening Sauce:
½ cup water
1 tsp sugar
2 tsp cornstarch
1 tsp soy sauce
1 tbsp oyster sauce

TIPS 貼士：

煎豆腐通常會用硬豆腐，其實用布包豆腐的好處是既軟滑而邊皮卻比滑豆腐硬，不輕易弄碎。只要煎豆腐時用猛鑊熱油，豆腐落鑊前確保吸乾水份，逐件落鑊，待煎至乾身才反轉，並添加少許油繼續煎，這樣便會煎出金黃軟滑的效果。如果炸豆腐，味道會更好。

People usually use hard tofu for pan-frying. Actually, it tastes even better to use wrapped soft tofu for its tenderness and its firmer outer part than soft tofu. Remember to heat up the pan and oil before putting in pat-dried tofu pieces one by one. Pan-fry one side of tofu until dry and brown before turning it over. Add in some more oil to continue pan-frying the other side to make the whole piece golden and soft. You may also deep-fry the tofu for a tastier flavour.

做法：

1. 豬肉切片加醃料拌勻。

2. 蔥切粒；草菇開半；青豆用滾水浸過備用。

3. 豆腐切厚件，灑少許鹽調味。

4. 燒熱鑊下2湯匙油，豆腐落鑊煎香，盛起。

5. 燒1湯匙油，下豬肉爆香，加薑、蒜、草菇、青豆炒透，潷酒，注入芡汁煮滾，豆腐可回鑊煮至入味，上碟，灑上蔥粒裝飾。

Method:

1. Cut pork into slices and then mix with marinade thoroughly.

2. Cut spring onion into dices; cut straw mushrooms in half; soak green peas in boiling water and then set aside.

3. Cut tofu into thick pieces, sprinkle them a little salt to taste.

4. Add 2 tbsp of oil into a hot frying pan, pan-fry tofu in it until golden brown, and then transfer it to dish.

5. Heat 1 tbsp of oil, stir-fry pork in it for a while, add in ginger, garlic, straw mushrooms and green peas and stir thoroughly. Splash in some wine, pour in thickening sauce and bring it to a boil. Return the tofu to the pan and cook until it becomes tasty. Transfer to a dish and then sprinkle some spring onion dices on top as garnishing.

雞粒扒玉子蛋豆腐

Braised Chicken with
Egg Tofu

預備時間：**10 分鐘** ▉烹調時間：**10 分鐘** ▉份量：**4 人**

▉ Preparation time：**10 mins**
▉ Cooking time：**10 mins**
▉ Serves：**4**

材料：

玉子豆腐2條
大雞蛋1隻
雞柳2兩 (80克)
雜豆4湯匙
靈芝菇1盒
乾葱碎1/2湯匙

醃料：

鹽1/8茶匙
糖1/8茶匙
生抽1/2茶匙
生粉1/2茶匙
水1湯匙

調味：

水3/8杯
鹽1/4茶匙
胡椒粉少許

芡汁：

水1/2杯、鹽1/8茶匙
糖1/8茶匙、生抽1茶匙
生粉1/2湯匙、麻油少許

Ingredients:

2 pcs egg tofu
1 large egg
80g chicken fillets
4 tbsp assorted peas
1 packet lingzhi mushrooms
1/2 tbsp minced shallots

Marinade Ingredients:

1/8 tsp salt
1/8 tsp sugar
1/2 tsp soy sauce
1/2 tsp cornstarch
1 tbsp water

Seasonings:

3/8 cup water
1/4 tsp salt
A pinch of pepper

Thickening Sauce:

1/2 cup water
1/8 tsp salt
1/8 tsp sugar
1 tsp soy sauce
1/2 tsp cornstarch
A little sesame oil

TIPS 貼士:

如果想吃蒸滑蛋但又不方便吃太多雞蛋，這個菜是最合適了。只用了一隻雞蛋便有一碟蒸滑蛋的效果，因玉子豆腐的口感與滑蛋差不多，再加入雞柳及鮮菇，營養豐富，老少咸宜。

This dish is just right for those who love steamed egg but cannot eat too much of it. By using egg tofu with texture similar to scrambled egg, it only takes one egg to make a dish which tastes like steamed egg. Moreover, the addition of chicken fillet and fresh mushrooms enriches its nutrition and makes it suitable for all.

做法：

1. 每條玉子豆腐切成八件厚片，排碟上。

2. 雞蛋打散，加調味料拌勻，用筲箕過濾注入玉子豆腐內，隔水以中慢火蒸約6分鐘，焗火備用。

3. 雞柳切粒，加醃料拌勻。

4. 靈芝菇去菇腳沖洗乾淨；分別與雜豆拖水，盛起。

5. 燒熱油，下乾葱碎及雞粒炒香，加靈芝菇及雜豆炒透，潷酒，加芡汁煮滾，鋪玉子蛋豆腐面，即成。

Method:

1. Cut each piece of egg tofu into 8 thick slices and then line them on a plate.

2. Beat the egg, add in seasonings and mix well. Filter the mixture with a sieve before adding it to egg tofu. Steam the plate of tofu above water on medium to low heat for about 6 mins, turn off the heat and then set aside.

3. Cut chicken fillets into dices, add in marinade and mix thoroughly.

4. Remove the stems of lingzhi mushrooms and then rinse with water. Blanch with assorted peas in boiling water and then take them out.

5. Heat up some oil, add in shallots and chicken dices, and then stir-fry for a while. Add in lingzhi mushrooms and assorted peas, and stir-fry. Splash in some wine, add in thickening sauce and bring it to a boil. Spread the mixture on egg tofu and serve.

八寶雜菜牛肉湯

Beef Soup with Assorted Vegetables

預備時間：**15分鐘** ■烹調時間：**15分鐘** ■份量：**4人**
Preparation time：**15 mins**
Cooking time：**15 mins**
Serves：**4**

材料：

免治牛肉4兩 (150克)
洋葱¼個
薯仔1個
甘筍仔1條
番茄3個
細V8蔬菜汁1罐
清水5杯

醃料：

糖½茶匙
生抽2茶匙
生粉1茶匙
1湯匙水
椒粉、麻油少許胡
1湯匙油 (後下)

調味：

鹽½茶匙
胡椒粉少許

Ingredients:
150g minced beef
¼ onion
1 potato
1 carrot
3 tomatoes
1 V8 vegetable juice (small can)
5 cups water

Marinade Ingredients:
½ tsp sugar
2 tsp soy sauce
1 tsp cornstarch
1 tbsp water
A little pepper and sesame oil
1 tbsp oil (to be added later)

Seasonings:
½ tsp salt
A pinch of pepper

◥ TIPS 貼士：

這個牛肉湯與家常做法不同，因為蔬菜是刨絲切碎，再加入含有多種元素的蔬菜汁，令牛肉湯更香濃更營養豐富。有人說這是西餐湯，無論如何定義，總之這是很受小朋友歡迎的濃湯，平時可作湯飯或湯通粉做午餐。

The making of this beef soup is quite different from the ordinary one as the vegetables are shredded with the addition of vegetable juice containing various elements, enriching the flavour and nutrition of the soup. Some people think it is a western soup but no matter what, it is a popular thick soup amongst children and can be cooked with rice or macaroni for lunch.

做法：

1. 牛肉加醃料拌勻，待醃15分鐘。
2. 番茄去皮切粒；洋葱切粒；甘筍、薯仔去皮刨絲。
3. 燒熱鍋，加1湯匙油爆香洋葱、甘筍、薯絲，注入清水煮10分鐘，加入牛肉碎及番茄粒煮滾，最後加蔬菜汁翻滾，下調味便可上碗。

Method:

1. Mix beef with marinade thoroughly and then leave it for 15 mins.
2. Peel tomatoes and then cut into dices; cut onion into dices; peel carrot and potato and then cut them into shreds.
3. Heat up a pot, add in 1 tbsp of oil, stir-fry onion, carrot and potato for a while. Add in water and cook for 10 mins. Add in minced beef and tomatoes, bring it to a boil, and finally add in vegetable juice and cook until it boils again. Add seasonings to taste. Pour it into a bowl and serve.

雪花海鮮羹
Seafood and Beancurd Soup

預備時間：**15分鐘** ■烹調時間：**10分鐘** ■份量：**4人**

■ Preparation time：**15 mins**
■ Cooking time：**10 mins**
■ Serves：**4**

材料：
蝦肉2兩 (80克)
魚肉2兩 (80克)
蟹柳3條
蟹味菇½盒
芫荽1棵
淡味豆腐花1杯
清雞湯2杯
水2杯

調味：
鹽¼茶匙
生粉¼茶匙
胡椒粉少許

芡汁：
生粉½湯匙
水3湯匙
鹽¼茶匙
胡椒粉、麻油少許

Ingredients:
80g shrimp meat
80g fish fillet
3 sticks Japanese crab meat
½ packet crab mushrooms
1 sprig coriander
1 cup un-sweeten tofu fa (beancurd jelly)
2 cups chicken broth
2 cups water

Seasonings:
¼ tsp salt
¼ tsp cornstarch
A pinch of pepper

Thickening Sauce:
½ tbsp of cornstarch
3 tbsp water
¼ tsp salt
A little pepper and sesame oil

TIPS 貼士：

用豆腐花做海鮮羹鮮甜香滑，若買不到豆腐花的話，可改用盒裝滑豆腐，口感與桶裝豆腐花差不多，切粒加入海鮮羹便可，若能加入鮮蟹肉會更加美味。

Tofu fa (beancurd jelly) in barrel sold in soy bean products stores is most suited for this soup with its refreshing flavour and smoothness. If not available, you may substitute it with soft tofu in packet. Simply cut tofu into dices before adding it into the soup. You may also add in crab meat to enhance its taste.

做法：
1. 蝦肉去腸，洗淨；魚肉切粒，分別加調味料拌勻。
2. 芫荽切碎；蟹柳切粒；蟹味菇剪去菇腳，沖水。
3. 燒半鑊滾水，分別把海鮮及蟹味菇略拖水半分鐘，盛起。
4. 用鑊燒少許油，下蟹味菇炒香，灒酒，倒入雞清湯及水煮滾，加海鮮料及豆腐花翻滾，埋芡汁煮至稠身，灑上芫荽碎，上窩。

Method:
1. Remove intestines of shrimps and then wash thoroughly; cut fish fillets into dices; mix them with seasonings separately.
2. Chop coriander finely; cut crab meat sticks into dices; remove stems of crab mushrooms and rinse with water.
3. Heat up half wok of water, blanch seafood and crab mushrooms in it for half a minute separately, take them out.
4. Heat up a little oil in a pot, add in crab mushrooms and stir-fry. Splash in some wine, pour in chicken broth and water, and bring it to a boil. Add in seafood and tofu fa, and bring it to a boil again. Stir in thickening sauce and cook until it thickens. Sprinkle chopped coriander on top. Transfer to a tureen and serve.

瓜菜

上湯浸釀素翅瓜甫

Braised Vegetarian Shark Fins and Stuffed Melons

預備時間：**30 分鐘** ■烹調時間：**15 分鐘** ■份量：**4 人**

■ Preparation time：**30 mins**
■ Cooking time：**15 mins**
■ Serves：**4**

材料：
免治豬肉4兩 (150克)
冬菇1隻，浸軟
大地魚末2茶匙
節瓜2個 (直身)
素濕翅4湯匙
清雞湯1杯

醃料 (肉)：
鹽及糖各½茶匙
生抽1茶匙
生粉½湯匙
胡椒粉、麻油少許

芡汁：
水¼杯
生粉1茶匙
麻油少許

Ingredients:
150 minced pork
1 shiitake mushroom (soaked)
2 tsp dried bonito fish flakes
2 hairy gourds (stiff in shape)
4 tbsp vegetarian shark fins
1 cup chicken broth

Marinade Ingredients (for meat) :
½ tsp salt and ½ tsp sugar
1 tsp soy sauce
½ tbsp cornstarch
A little pepper and sesame oil

Thickening Sauce:
¼ cup water
1 tsp cornstarch
A little sesame oil

TIPS 貼士：

雜貨舖買回來的大地魚，不用洗，把魚肉撕下來，魚皮也要撕走，淨肉剪碎，下4湯匙油內炸至香脆，取出，可用機磨碎或用刀尾椿碎，便成大地魚末。加入豬肉內美味無窮。釀節瓜甫也可用蒸的方法，不過利用這樣的煎煮法，節瓜會更加入味。

Dried bonito flakes is made from dried bonito available in grocery stores. Bone the fish without washing it, tear off the skin from the fillet, cut fish meat into small pieces and then deep-fry it with 4 tbsp of oil until crispy. Take it out, grind it finely with a food processor or smash it with the handle of knife and then fold in pork to create a wonderful flavour. Stuffed hairy gourd can also be cooked by steaming but this pan-frying and braising method is recommended for a better taste.

做法：

1. 冬菇浸軟切幼粒；大地魚肉用油炸香，磨碎成木狀。

2. 豬肉加醃料攪拌至起膠，再加冬菇粒、大地魚末拌勻。

3. 素翅用熱水沖洗乾淨，隔水備用。

4. 節瓜刮去毛，修切成8-10件約1.5厘米厚件，中央挖半空，並上少許生粉，釀入豬肉，再鋪上素翅裝飾。

5. 用平底鑊燒1湯匙油，放下瓜甫略煎，注入雞清湯，蓋好慢火焗煮10分鐘或至熟透，瓜甫排碟上，餘卜汁液加芡汁煮至汁稠身，淋瓜甫面，即成。

Method:

1. Soak shiitake mushrooms until softened and then cut into fine dices; deep-fry bonito fish fillet in oil until aromatic and then ground into powder.

2. Stir pork with marinade until sticky, add in mushroom dices and bonito fish powder, and blend well.

3. Rinse vegetarian shark fins with hot water thoroughly, strain and then set aside.

4. Scrape the hair of hairy gourds, cut into 8-10 pcs of about 1.5 cm thick, scoop out the core in the centre, smear with a little cornstarch, stuff in minced pork, and then spread some vegetarian shark fins on top to garnish.

5. Heat up 1 tsp of oil in a frying pan, add in gourd pieces to pan-fry slightly, add in chicken broth, cover the pan with a lid and simmer on low heat for 10 mins until cooked. Line the gourd rings on a plate; add thickening sauce to the remaining sauce and cook until it thickens, pour it over the gourd rings and serve.

松仁素肉生菜包

Vegetarian Ham with Pine Nuts in Lettuce

預備時間：**20 分鐘** ■烹調時間：**10 分鐘** ■份量：**4 人**

■ Preparation time：**20 mins**

Cooking time：**10 mins**

Serves：**4**

材料：

素火腿2兩 (80克)

紅蘿蔔1兩 (40克)

西芹2兩 (80克)

青椒½隻

紅辣椒½隻

甜菜脯2條

冬菇2隻

松子仁2湯匙

生菜數片

海鮮醬適量

芡汁：

水1湯匙

鹽⅛茶匙

糖⅛茶匙

生抽1茶匙

生粉½茶匙

Ingredients:

80g vegetarian ham

40g carrot

80g celery

½ green pepper

½ red chili

2 pcs preserved sweetened turnips

2 shiitake mushrooms

2 tbsp Pine nuts

A few lettuce leaves

Some Hoisin sauce

Thickening Sauce:

1 tbsp water

⅛ tsp salt

⅛ tsp sugar

1 tsp soy sauce

½ tsp cornstarch

⬊ TIPS 貼士：

素食者當然要用素火腿，可在凍肉舖買到。炒雜菜粒用的蔬菜隨意選擇，只要夠爽脆，色澤夠吸引，最重要的精髓是用甜菜脯，它令素菜顯得惹味，除了同生菜包同吃外，也可用炸蝦片伴吃。

Vegetarian ham is available in frozen food stores. You may choose whatever vegetables you like, as long as they are crunchy and colourful, for making stir-fried assorted vegetables. However, you must not miss out the sweetened preserved turnips, which plays a key role in bringing out the flavour of the vegetables. Instead of using lettuce, you may serve this vegetables with prawn crackers.

做法：

1. 冬菇浸軟去蒂，加少許酒，糖及生粉拌勻，隔水先蒸10分鐘，取出，切粒。

2. 甜菜脯浸水後切粒；素火腿及其餘蔬菜切粒；生菜洗淨修剪好備用。

3. 燒熱鑊加2湯匙油，下紅蘿蔔、冬菇及甜菜脯先炒香，再加西芹及其餘材料炒透，加入芡汁兜至乾身，上碟，灑上松子仁，食時用生菜及海鮮醬伴食，便成健康美味生菜包。

Method:

1. Soak shiitake mushrooms until softened and then remove their stems. Add in a little wine, sugar and cornstarch and mix well. Steam above water for 10 mins, take it out and then cut into dices.

2. Cut preserved turnips in water and then cut into dices; cut vegetarian ham and the rest of the vegetables into dices, trim lettuce leaves and then set aside.

3. Heat up a wok, add in 2 tbsp of oil, carrot, mushrooms and preserved turnips, and stir well. Add in celery and the remaining ingredients, and stir fry thoroughly. Stir in thickening sauce and cook until the sauce reduces. Transfer in a dish, sprinkle some pine nuts on top, wrap it with lettuce leaves and then add some Hoisin sauce to taste.

洋葱牛肉薯絲餅

Onion and **Beef**
Hash Browns

預備時間：**15分鐘** ■烹調時間：**10分鐘** ■份量：**4人**

■ Preparation time：**15 mins**

■ Cooking time：**10 mins**

Serves：**4**

材料：
免治牛肉3兩（120克）
洋葱1/2個
薯仔絲8兩（300克）
番茄1個
茄汁適量

醃料：
糖1/2茶匙
生粉2茶匙
生抽2茶匙
黑胡椒碎1/4茶匙
蛋黃1隻

調味：
鹽1/2茶匙
麵粉4-5湯匙

Ingredients:
120g minced beef
1/2 onion
300g potato shreds
1 tomato
Some ketchup

Marinade Ingredients:
1/2 tsp sugar
2 tsp cornstarch
2 tsp soy sauce
1/4 tsp black pepper
1 egg yolk

Seasonings:
1/2 tsp salt
4-5 tbsp flour

⬂TIPS 貼士：

薯仔刨絲後要吸乾水，與牛肉撈勻後要即時落鑊煎香，否則薯絲會滲出水，煎出的薯餅就不夠香脆。轉用豬肉或雞肉也可，但煎時要蓋上鑊蓋焗片刻，好讓肉碎熟透，開蓋後再煎至脆身便可。

The shredded potatoes must be patted dry and then pan-fried immediately after mixing with beef, otherwise, water will seep from the potatoes and make the hash brown less crispy. If you replace beef with pork or chicken fillet, you must cook it with the wok covered with a lid for a while to allow the meat to be fully cooked, and then remove the lid to pan-fry until crispy.

做法：

1. 免治牛肉加醃料拌勻醃片刻；洋葱切幼粒，加牛肉內拌勻。

2. 番茄切片備用；薯仔去皮刨粗絲，浸水後搾乾水，即灑下調味料再與牛肉拌勻。

3. 鑊內燒3湯匙油，放一湯匙薯絲至熱油內以中火煎香，反轉另一面略壓扁，繼續煎至薯餅金黃熟透，上碟。以番茄片伴碟及跟茄汁享用，更受小朋友歡迎。

Method:

1. Mix minced beef with marinade thoroughly and then leave it for a while; cut onion into fine dices, add in beef and mix well.

2. Cut tomato into slices and then set aside. Peel potato and then shred, soak in water and then squeeze to dry, sprinkle with seasonings and then mix it with beef.

3. Heat up 3 tbsp of oil in a frying pan, put in ladles of shredded potato one by one to pan-fry on medium heat until brownish, flip over and press gently to flatten, pan-fry until browned and cooked, and then transfer to a plate. Serve with tomato slices and ketchup. This is a popular dish to children.

馬拉盞雞絲炒通菜

Stir-fried Water Spinach
with Chicken in Spicy Sauce

預備時間：**15分鐘** ▧烹調時間：**10分鐘** ▧份量：**4人**

▧ Preparation time：**15 mins**
▧ Cooking time：**10 mins**
▧ Serves：**4**

材料：

雞髀肉4兩 (150克)
通菜12兩 (450克)
辣椒仔2隻
薑片1片
蒜蓉1湯匙
蝦醬2湯匙

醃料：

糖1/4茶匙
生抽1茶匙
生粉1茶匙

芡汁：

水1湯匙
糖1/2茶匙
蠔油1/2茶匙
生粉1/2茶匙

Ingredients:

150g chicken thigh meat
450g water spinach
2 red chilies
1 slice ginger
1 tbsp minced garlic
2 tbsp shrimp paste

Marinade Ingredients:

1/4 tsp sugar
1 tsp soy sauce
1 tsp cornstarch

Thickening Sauce:

1 tbsp water
1/2 tsp sugar
1/2 tsp oyster sauce
1/2 tsp cornstarch

▧ TIPS 貼士：

酒樓食肆會用大量油炒通菜，通菜便不會轉瘀黑色，特別好味。不過在家中要炒靚通菜，首先要夠鑊氣，第一輪要將通菜用高溫炒焗至軟身，然後再起鑊爆醬料把通菜回鑊再炒，這樣經過兩次用油高溫兜炒，當然惹味好吃。

Restaurants use lots of oil to stir-fry water spinach so as to prevent it from turning dark green while making it extra delicious. If you want to make it well at home, you have to cook it with a very hot wok. First of the all, stir-fry the water spinach quickly and then cover it with a lid to cook on high heat until it becomes soft. Then stir-fry the sauce ingredients in the wok thoroughly before returning the vegetable to stir-fry again. By stir-frying on high heat twice, the water spinach will definitely turn into a mouth-watering dish.

做法：

1. 雞髀肉去皮切條，加醃料同醃。
2. 通菜洗淨，摘成長度；薑切絲；辣椒仔略拍扁。
3. 燒2湯匙油，下薑絲、半份蒜蓉、通菜及辣椒仔炒透，蓋好以大火煮1分鐘至軟身，隔乾水。
4. 燒熱鑊，加1湯匙油，下雞柳炒至八成熟，盛起。
5. 燒1湯匙油，慢火爆香蒜蓉、蝦醬，雞柳回鑊炒透，潷酒，通菜同鑊加芡汁兜勻，上碟。

Method:

1. Skin chicken thigh meat, cut into strips, and them mix with marinade.
2. Wash water spinach thoroughly, tear into sticks, shred ginger; slightly smash red chilies.
3. Heat up 2 tbsp of oil, add in ginger, half of minced garlic, water spinach and chilies, and stir-fry thoroughly. Cover with a lid and cook on high heat for 1 min until softened, and then strain to dry.
4. Heat up a wok, add in 1 tbsp of oil, chicken fillets and stir-fry until almost cooked, and then transfer to a dish.
5. Heat up 1 tbsp of oil, stir-fry minced garlic and shrimp paste on low heat, return chicken fillet and stir-fry, splash in some wine, return water spinach, add in thickening sauce and stir thoroughly. Transfer to a dish and serve.

蛋白蟹柳扒鮮菇

Braised Mushrooms
with Egg White and
Japanese Crab Meat

預備時間：**15分鐘** ■烹調時間：**10分鐘** ■份量：**4人**
■ Preparation time：**15 mins**
■ Cooking time：**10 mins**
■ Serves：**4**

材料：

鮮草菇12兩 (450克)
蟹柳6條
蛋白2隻
薑4片
葱1條

煨料：

水2杯
鹽及糖各¼茶匙
雞粉2茶匙
紹興酒1湯匙

芡汁：

清雞湯1杯
生粉2茶匙
胡椒粉、麻油少許

Ingredients:

450g fresh straw mushrooms
6 pcs Japanese Crab meat
2 egg white
4 slices ginger
1 sprig spring onion

Braising Sauce:

2 cups water
¼ tsp salt and ¼ tsp sugar
2 tsp chicken powder
1 tbsp Shaoxing wine

Thickening Sauce:

1 cups chicken broth
2 tsp cornstarch
A little pepper and sesame oil

TIPS 貼士：

鮮草菇要買乾身淺色的為佳，要用小刀修切頂部黑色部份，還要剠十字花，使其容易入味。它一定要出透水才能去除泥味，然後加煨料煨過。如果選用罐頭草菇也無妨，但也要做煨的步驟才會好吃。

Select the fresh straw mushrooms which are dry and light in colour. Trim out the dark portion on top and then cut a cross with a small knife to let the flavour permeate into it easily. Remember to blanch it in boiling water long enough to remove its muddy flavour before braising it with seasonings. You may use canned straw mushrooms but you must also follow the braising steps to make it tasty.

做法：

1. 草菇修切頂部及剠十字，放滾水內拖水5分鐘，取出，沖水。

2. 燒少許油爆香薑、葱，下草菇炒透，加煨料煮7分鐘，隔起。

3. 蛋白拌勻；蟹柳拆成絲。

4. 燒1湯匙油，下草菇炒透，潷酒，注入半份芡汁兜至入味，上碟。

5. 燒1湯匙油，加蟹柳及芡汁煮滾，拌入蛋白捹至汁濃稠，便可鋪草菇面。

Method:

1. Cut a cross on the top of straw mushrooms, blanch in boiling water for 5 mins, take them out and rinse with water.

2. Heat up a little oil, stir-fry ginger and spring onion, add in straw mushrooms and stir fry, add in braising sauce ingredients and cook for 7 mins, and then take it out.

3. Beat egg white thoroughly; tear Japanese crab meat into shreds.

4. Heat up 1 tbsp of oil, add in straw mushrooms and stir-fry thoroughly, splash in wine, add in half portion of thickening sauce and stir until tasty. Transfer to a dish.

5. Heat up 1 tbsp of oil, add in crab meat and remaining thickening sauce, and bring it to a boil. Stir in egg white and stir gently until thickens. Pour it over straw mushrooms and serve.

瑤柱扒釀竹笙筒

Braised Dried Scallops
with Dried Bamboo
Fungi Rings

預備時間：**30分鐘** ■烹調時間：**10分鐘** ■份量：**4人**

■ Preparation time：**30 mins**
■ Cooking time：**10 mins**
■ Serves：**4**

材料：

瑤柱½兩 (20克)
雞清湯5湯匙
竹笙12條
熟火腿條12條
紅蘿蔔條12條
西蘭花梗條12條
大冬菇2隻

煨料：
雞清湯½杯
清水½杯
紹興酒1茶匙

芡汁：
雞清湯½杯
蠔油1茶匙
生粉1½茶匙
麻油數滴

Ingredients:

20g dried scallops
5 tbsp chicken broth
12 pcs dried bamboo fungi (zhusheng)
12 pcs cooked ham
12 carrots
12 broccoli strips
2 large shiitake mushrooms

Braising Sauce:
½ cup chicken broth
½ cup water
1 tsp Shaoxing wine

Thickening Sauce:
½ cup chicken broth
1 tsp oyster sauce
1½ tsp cornstarch
A few drops of sesame oil

⬊ TIPS 貼士：

竹笙有兩種貨質，一種是包裝好的，較薄身，包裝裏有很多"鬚根"，價錢很便宜，所以煮湯也無妨。另一種是散買的，以兩計算，肉身較厚，口感較爽口，屬佳品，適宜釀。竹笙其實無味，要出水後再用上湯煨過才會好味。

There are two types of bamboo furgi, one in packet and the other is sold by weight. The packet one is thin with lots of "roots" but cheap in price, and is suitable for making soup. The other one is thicker with a crunchier texture, and is most suitable for making stuffed zhusheng rings. Bamboo furgi is bland in taste and so it has to be blanched in boiling water and then braised in stock.

做法：

1. 瑤柱加雞清湯浸軟，隔水蒸20分鐘，取出，隔去瑤柱水備用；瑤柱拆成絲。
2. 冬菇浸軟切條；竹笙修剪好頭尾，洗淨浸軟，再拖水，壓乾水。
3. 煨料加冬菇及竹笙煮滾，熄火，待浸至入味，隔去汁液，備用。
4. 西蘭花梗去硬皮後切條；火腿及紅蘿蔔切成相若幼條。
5. 把每條竹笙釀入冬菇、火腿、紅蘿蔔及西蘭花梗各一條，排放碟上，隔水蒸約6分鐘，取出。
6. 燒1湯匙油，爆香瑤柱絲，潷酒，加芡汁及瑤柱水煮滾，淋在竹笙筒上。

Method:

1. Soak dried scallops in chicken broth until softened, steam above water for 20 mins, take it out, strain and then set aside; tear into shreds.
2. Soak shiitake mushrooms until softened and then cut into strips, trim the ends of dried bamboo fungi, wash thoroughly and soak until softened, blanch in hot water, and then press to dry
3. Cook shiitake mushrooms and dried bamboo fungi with braising sauce, bring it to a boil, turn off the heat, and leave it until flavour permeates into the food. Filter out the sauce and then set aside.
4. Peel broccoli steam and then cut into strips; cut ham and carrots into fine strips of similar size.
5. Stuff shiitake mushrooms, ham, a carrot strip and a broccoli strip into a dried bamboo fungi ring. Line the stuffed fungi on a plate, steam above water for about 6 mins and then take it out.
6. Heat up 1 tbsp of oil, stir-fry dried scallop shreds, splash in some wine, add in thickening sauce and dried scallop water, and bring it to a boil. Pour it over the dried bamboo fungi and serve.

瑤柱金菇扒菠菜

**Braised Dried Scallops
and Enoki Mushrooms
with Spinach**

預備時間：**20 分鐘** ■烹調時間：**10 分鐘** ■份量：**4 人**
■ Preparation time：**20 mins**
■ Cooking time：**10 mins**
■ Serves：**4**

材料：	Ingredients:
菠菜12兩 (450克)	450g spinach
金菇1包	1 packet of enoki mushrooms
瑤柱½兩 (20克)	20g dried scallops
蒜頭2粒	2 garlics
薑米1茶匙	1 tsp minced ginger

調味 (A)： / **Seasonings (A):**

水5湯匙	5 tbsp water
雞粉½茶匙	½ tsp chicken powder

調味 (B)： / **Seasonings (B):**

雞粉1茶匙	1 tsp chicken powder
糖½茶匙	½ tsp sugar

芡汁： / **Thickening Sauce:**

水½杯	½ cup water
糖¼茶匙	¼ tsp sugar
蠔油1湯匙	1 tbsp oyster sauce
生粉½湯匙	½ tbsp cornstarch
胡椒粉及麻油少許	A little pepper and sesame oil

⊠ TIPS 貼士：

瑤柱要浸水蒸軟才拆絲。如果想快捷，也可將瑤柱加水及調味放入微波爐用高火煮4-5分鐘，夠軟身後便可拆絲。加入金菇目的是減少用瑤柱份量，既經濟，又健康。

Dried scallops have to be soaked and then steamed to soften before shredding. For an easier method, you may cook the dried scallops with water and seasonings on high heat in microwave oven for 4-5 minutes to make it soft enough for shredding. The addition of enoki mushrooms can reduce the amount of dried scallops used and thus making it more economical and healthier.

做法：

1. 瑤柱用調味 (A) 先浸軟，再隔水蒸20分鐘，隔去瑤柱水備用，瑤柱用菜刀壓扁拆成絲。

2. 金菇切去菇腳，開半切成短度；蒜頭拍扁；薑切幼粒；菠菜切短度。

3. 燒2湯匙油，爆香蒜頭，加菠菜及調味 (B) 蓋好煮1分鐘，開蓋兜至軟身，隔乾水後放碟上。

4. 燒熱鑊，下1茶匙油爆香薑米及金菇，潲酒兜勻，隔水備用。

5. 燒1湯匙油，加瑤柱爆香，金菇回鑊加芡汁及瑤柱水煮至濃稠，鋪菠菜面。

Method:

1. Soak dried scallops in seasonings (A), steam above water for 20 mins, filter out the dried scallop water and then set aside, tear dried scallops into shreds.

2. Cut the stems of enoki mushrooms, cut in half into short sticks, smash garlic lightly, cut ginger into dices; cut spinach into short sticks.

3. Heat up 2 tbsp of oil, stir-fry garlic, add in spinach and seasonings (B) and then cover with a lid to cook for 1 min. Remove the lid and stir-fry until softened. Strain to dry and then place on a dish.

4. Heat up a wok, add in 1 tsp of oil, stir-fry minced ginger and enoki mushrooms, splash in some wine and stir-fry, strain to dry and then set aside.

5. Heat up 1 tbsp of oil, add in dried scallops, return enoki mushrooms, add in thickening sauce and dried scallop water to cook until thickens, and then spread it over spinach.

腐乳冬瓜煮魚鬆

Braised Fish Balls with
Fermented Beancurd
and Winter Melon

預備時間：**10 分鐘** ■烹調時間：**20 分鐘** ■份量：**4 人**
■ Preparation time：**10 mins**
■ Cooking time：**20 mins**
■ Serves：**4**

材料：

鯪魚肉4兩 (150克)
陳皮末1茶匙
葱粒 1湯匙
冬瓜1斤 (600克)
薑3片
辣椒腐乳4塊
夜香花數朵 (隨意)

醃料(魚肉)：

鹽1/4茶匙
生粉2茶匙
水1 1/2湯匙
胡椒粉、麻油少許

調味：

水1杯
鹽1/4茶匙
糖1/4茶匙
胡椒粉少許

Ingredients:

150g minced mud carp meat
1 tsp dried tangerine peel (soaked and chopped)
1 tbsp spring onion dices
600g winter melon
3 slices ginger
4 cubes chili fermented beancurd
Some night-fragrant flower

Marinade Ingredients (for minced fish)：

1/4 tsp salt
2 tsp cornstarch
1 1/2 tbsp water
A little pepper and sesame oil

Seasonings:

1 cup water
1/4 tsp salt
1/4 tsp sugar
A pinch of pepper

◣TIPS 貼士：

在市場買到的鯪魚肉有兩種，一是已調味的，一是無調味的。最好買無調味的較新鮮，又可自己調較喜愛的味道，可加入陳皮末來辟除魚腥及泥味。下調味後要順一方向攪拌至起膠，鯪魚膠方可加入其他配料，如芫荽、葱或火腿也可，煮食方法可煎、炸、蒸或煮，各有特色，方便好用。

There are two types of minced mud carp meat available, one is mixed with seasonings and the other is not. It is best to buy the one without seasonings which is not only fresher but also easily adjusted to one's taste. You may also mix it with dried tangerine to remove its fishy smell and mud flavour. When folding in seasonings, you must stir it unilaterally until it becomes sticky. You may then add in other ingredients, such as coriander, spring onion and ham, and then cook it by pan-frying, deep frying, steaming or poaching for a different taste.

做法：

1. 陳皮浸軟挖去瓤，剁成幼末；冬瓜去皮切厚件；夜香花洗淨備用。

2. 鯪魚肉加醃料攪成膠狀，再加陳皮末及葱粒拌勻。

3. 用鑊燒1杯滾油，用湯匙撥下一粒粒魚膠，炸成金黃鯪魚球狀，隔油。

4. 燒1湯匙油，下薑片及腐乳爆香，加冬瓜件兜勻，注入調味蓋奅煮6-8分鐘至瓜軟身，魚球回鑊煮至入味，最後拌入夜香花煮滾，即成夏日消暑家常小菜。

Method:

1. Scrape the membrane of dried tangerine peels and then chop finely; peel winter melon and then cut into thick pieces; wash spearmint and then set aside.

2. Mix minced mud carp meat with mariande, stir the mixture into a paste, add in dried tangerine peel choppings and spring onion dices, and mix well.

3. Heat up 1 cup of oil, put in small spoonful of fish paste, deep-fry Into fish balls, and then strain out excess oil.

4. Heat up 1 tbsp of oil, add in ginger slices and fermented beancurd, add in winter melon and stir-fry thoroughly. Add in seasonings, cover it with a lid and cook for 6-8 mins until the melon is soft. Return fish balls to the wok and cook until tasty. Finally, add in night fragrant flower, bring it to a boil and ready to serve. This is light homemade dish in summer.

鐵板煎釀茄子

Stuffed Eggplant
Teppanyaki

預備時間：**15 分鐘**　烹調時間：**10 分鐘**　■份量：**4 人**
■ Preparation time：**15 mins**
■ Cooking time：**10 mins**
■ Serves：**4**

材料：

免治豬肉2兩 (80克)

鯪魚肉4兩 (150克)

茄子1大條

薑米1茶匙

蔥1條

和風照燒汁3湯匙

炒香芝麻1湯匙

醃料：

鹽1/2茶匙

糖1/8茶匙

生抽1/2茶匙

生粉1/2湯匙

水1湯匙

胡椒粉、麻油少許

Ingredients:
80g minced pork
150g minced mud carp meat
1 large eggplant
1 tsp minced ginger
1 sprig spring onion
3 tbsp terriyaki dressing
1 tbsp fried sesame seed

Marinade Ingredients:
½ tsp salt
⅛ tsp sugar
½ tsp soy sauce
½ tbsp cornstarch
1 tbsp water
A little pepper and sesame oil

TIPS 貼士：

用鐵板上桌很有熱騰騰的感覺，預備鐵板時要注意，若要燒得夠熱，應將鐵板先加水煮至水完全蒸發，然後再乾燒片刻，直至鐵板熱透心，這時候當醬汁淋食物面時便會散發出濃烈香味，引發食慾。

Serving with a teppan can add a sizzling touch to the dish. To get it right, you must first boil some water in the teppan until liquid completely evaporates, and then heat it for a while until the teppan is thoroughly hot. When you pour the sauce over the food on teppan, the strong aroma will burst out to stimulate the one's appetite.

做法：

1. 免治豬肉與鯪魚肉同放大碗內加醃料順一方向攪拌至起膠。
2. 蔥切絲；茄子修切成12條長條，每條瓜肉中央切去少許瓜瓤，浸水備用。
3. 茄子吸乾水，塗少許生粉，釀上適量肉蓉，修至平滑。
4. 燒1杯滾油，魚肉向下，把釀茄子半煎炸至金黃色及熟透，盛起，吸乾油份。
5. 燒熱鐵板或平底鑊，卜薑米及釀茄子放鐵板上，鋪上蔥絲，澆上和風照燒汁，灑上芝麻裝飾，趁熱上桌。

Method:

1. Place minced pork and minced mud carp meat in a mixing bowl, add in marinade, and then stir unilaterally until sticky.
2. Shred spring onion, cut eggplant into 12 long strips, scrap the seeds and core, soak in water and then set aside.
3. Pat dry eggplant, smear with a little cornstarch, stuff in some minced meat paste, and then trim its side to smooth out.
4. Heat up 1 cup of oil, add in stuffed eggplant with minced meat paste, facing downs to deep-fry until golden brown and cooked. Take them out and then pat with oil absorbent paper.
5. Heat up a teppanyaki grill or frying pan, add in minced ginger and stuffed eggplant on the grill, spread spring onion shreds on top, pour some terriyaki sauce on top, sprinkle with sesame seed to garnish, and then serve hot.

欖菜肉碎炒翠玉瓜

Stir-fried Angle Luffa
with Preserved Cabbage
in Olive

預備時間：**10 分鐘** ■烹調時間：**10 分鐘** ■份量：**4 人**
■ Preparation time：**10 mins**
■ Cooking time：**10 mins**
■ Serves：**4**

材料：

免治豬肉2½兩（100克）
翠玉瓜12兩（450克）
薑2片
蒜蓉2茶匙匙
豆瓣醬½茶匙
橄欖菜2湯匙

醃料：

鹽¼茶匙
糖⅛茶匙
生抽1茶匙
生粉1茶匙
水1湯匙

調味：

水5湯匙
鹽¼茶匙
胡椒粉少許

芡汁：

水3湯匙
生抽½茶匙
糖¼茶匙
生粉½茶匙

Ingredients:

100g minced pork
450g angle luffa
2 slices ginger
2 tsp minced garlic
½ tsp chili bean sauce
2 tbsp preserved cabbage in olive

Marinade Ingredients:

¼ tsp salt
⅛ tsp sugar
1 tsp soy sauce
1 tsp cornstarch
1 tbsp water

Seasonings:

5 tbsp water
¼ tsp salt
A pinch of pepper

Thickening Sauce:

3 tbsp water
½ tsp soy sauce
¼ tsp sugar
½ tsp cornstarch

↖ TIPS 貼士：

和欖菜肉碎同炒的蔬菜最緊
要夠爽脆，常用有四季豆、
涼瓜和西蘭花，翠玉瓜也很
合適。要保持爽脆口感，翠
玉瓜不可切片，可將翠玉瓜
直切成四件，還要把瓜瓤切
走，然後斜切件。橄欖菜是
油浸鹹菜，切記要將油隔
淨，否則整碟菜會變得很
油膩。

Stir-fried preserved cabbage
and minced meat should be
accompanied by crunchy
vegetables, such as , snap
beans, bitter melon, broccoli
and angle luffa. However,
never cut angle luffa into
slices so as to preserve its
crunchiness. Simply cut it
vertically into 4 pieces, scrape
out its seeds and then cut
into wedges. Since preserved
cabbage is a salted vegetable
in oil, you must filter out its
excess oil before cooking or
your dish will be spoilt by its
grease.

做法：

1. 免治豬肉加醃料拌勻。

2. 翠玉瓜斜切厚片；薑切絲，橄欖菜隔去油份。

3. 燒1湯匙油，爆香薑絲及翠玉瓜，加調味料蓋好煮2分鐘至瓜半熟，盛起，隔去水份。

4. 燒1湯匙油，下肉碎及蒜蓉爆至十成熟，加豆瓣醬及橄欖菜兜炒，翠玉瓜回鑊炒過，加入芡汁兜炒至
   乾身，上碟。

Method:

1. Mix minced pork with marinade thoroughly.

2. Cut angle luffa diagonally into thick slices; cut ginger into shreds; filter out the oil from preserved cabbage in olive.

3. Heat up 1 tbsp of oil, stir-fry ginger shreds and angle luffa, add in seasonings, cover it with a lid and cook for 2 mins until medium cooked, take them out and then strain.

4. Heat up 1 tbsp of oil, add in minced pork and minced garlic to stir-fry until almost cooked, add in chili bean sauce and preserved cabbage. Return angle luffa to the wok and stir-fry thoroughly, add in thickening sauce until liquid reduces, and then transfer to a dish to serve.

百花紫菜蛋卷

Shrimp and
Seaweed Egg Rolls

預備時間：**30 分鐘** ■烹調時間：**15 分鐘** ■份量：**4 人**

■ Preparation time：**30 mins**

■ Cooking time：**15 mins**

■ Serves：**4**

材料：
中蝦肉5兩 (200克)
雞蛋2隻
雜豆3湯匙
壽司紫菜1塊
西蘭花1棵

醃料：
鹽1/3茶匙
生粉1/2湯匙
蛋白1茶匙
胡椒粉少許

調味：
鹽1/2茶匙
糖1/4茶匙
油1/2湯匙

芡汁：
清雞湯1/3杯
生粉1茶匙

Ingredients:
200g medium-size shrimps meat
2 eggs
3 tbsp assorted peas
1 pc sushi roll seaweed
1 broccoli

Marinade Ingredients:
1/3 tsp salt
1/2 tbsp cornstarch
1 tsp egg white
A pinch of pepper

Seasonings:
1/2 tsp salt
1/4 tsp sugar
1/2 tbsp oil

Thickening Sauce:
1/3 cup chicken broth
1 tsp cornstarch

◤TIPS 貼士:

一斤蝦去殼後可得八兩肉 (約300克)，蝦肉洗淨後緊記要吸乾水，拍蝦膠後可用膠撻順一方向用力攪拌及壓平，這樣處理蝦膠會不易鬆散。雪蝦膠時間越長會變得更爽口彈牙，未夠熟的蝦膠會有霉臉口感，所以要注意烹煮時間。

Normally, you can only get 300g of shrimp meat out of 600g of shrimps with shells. Remember to pat dry the meat after washing. Then smack the shrimps before stirring them together unilaterally with a spatula vigorously to blend into a firm paste. Afterwards, chill it in the refrigerator to give it an elastic texture: the longer the chilling, the better the result. When it is used for cooking, make sure it is cooked thoroughly or the shrimp paste will become a mash.

做法：

1. 蝦肉挑去黑腸，洗淨，吸乾水，用刀拍扁，再用刀背略剁成蝦蓉，加醃料順一方向攪成蝦膠狀，放冰箱內冷藏片刻。

2. 雞蛋加少許鹽拌勻；用少許油下蛋汁煎成一大片薄蛋皮，盛起。

3. 雜豆用滾水浸過，吸乾水，加1茶匙生粉拌勻。

4. 蝦膠放膠紙上壓成一大片與紫菜相若大方塊，蛋皮放底，紫菜放面，蝦膠鋪紫菜面，撕去膠紙，雜豆分佈在蝦膠上裝飾，捲成壽司卷型，放蒸架上隔水以中火蒸約13分鐘，取出，切成9件，砌放碟上。

5. 西蘭花修切好，放滾水內加調味料拖熟，盛起，排碟邊。

6. 燒少許油，下芡汁煮滾，淋面，趁熱享用。

Method:

1. Remove intestine from shrimps, wash thoroughly, pat dry, smash to flatten with knife blade, and then chop slightly with the back of the knife into shrimp paste. Add in marinade ingredients, stir unilaterally into a glue, and then chill in the refrigerator for a while.

2. Add a pinch of salt to eggs and beat thoroughly, pan-fry the egg liquid with a little oil into a large egg sheet.

3. Soak assorted peas in boiling water for a while, pat dry, add in 1 tsp of cornstarch and mix well.

4. Place shrimp paste on a cellophane paper and then press into a large flat sheet of size similar to that of a seaweed sheet. Place egg sheet on the base, seaweed on top and then place shrimp paste on seaweed top. Remove the cellophane paper, spread assorted peas on shrimp paste as garnishing, and then roll it up into a sushi form. Place it on a steaming rack to steam above water on medium heat for 13 mins. Take it out, cut into 9 pieces and then line them on a plate.

5. Trim broccoli, boil it in boiling water with seasonings until cooked, take it out, and then line it on the side of the dish.

6. Heat up a little oil, add in thickening sauce, bring it to a boil, and then pour it over the dish. Serve hot.

芒果酸辣蝦球

Sour and ~~Spicy Prawns~~
with Mango

預備時間：**15分鐘** ■烹調時間：**10分鐘** ■份量：**4人**
■ Preparation time：**15 mins**
■ Cooking time：**10 mins**
■ Serves：**4**

材料：

中蝦12隻
芒果2個
蒜頭1粒
青椒粒1湯匙
雞蛋1隻
生粉1/4杯
新鮮麵包糠1杯

醃料：

鹽1/3茶匙
蛋白1茶匙
生粉1茶匙
胡椒粉少許

芡汁：

泰國酸辣雞醬4湯匙
茄汁1湯匙
水1湯匙

Ingredients:

12 medium size shrimps
2 mangoes
1 garlic
1 tbsp green pepper dices
1 egg
1/4 cup cornstarch
1 cup fresh breadcrumb

Marinade Ingredients:

1/3 tsp salt
1 tsp egg white
1 tsp cornstarch
A pinch of pepper

Thickening Sauce:

4 tbsp Thai sour and spicy chicken sauce
1 tbsp ketchup
1 tbsp water

做法：

1. 中蝦去殼留尾，蝦背剝開起雙飛，去黑腸，沖洗乾淨，吸乾水，下醃料拌勻，放冰箱內冷藏片刻。
2. 芒果開半，起出果肉；切成粗粒；蒜頭切片；青椒切粒。
3. 把蝦隻先沾上生粉，再上蛋汁，最後沾滿麵包糠，逐隻放入熱油內炸至金黃色，隔油。
4. 燒少許油，下蒜片、青椒粒及芡汁煮滾，炸蝦及芒果回鑊快手兜勻，上碟，趁熱品嚐。

Method:

1. Remove the shell of shrimp while keeping its tail intact, cut its back open, remove its intestine, rinse with water and then pat dry. Mix with marinade ingredients and then chill in the refrigerator for a while.
2. Cut mango in half, scoop out the flesh, cut the flesh into dices, cut garlic into slices, cut green pepper into dices.
3. Coat shrimps with cornstarch, soak with egg liquid and then coat with breadcrumb. Put the shrimps into a wok of hot oil one by one to deep-fry until golden brown and then strain out excess oil.
4. Heat up a little oil, put in garlic slices, green pepper dices and thickening sauce, and then bring it to a boil. Return fried shrimps and mango to wok, stir-fry and then transfer to a plate. Serve hot.

奶油大蟹伊麵

Braised E-fu Noodles
with Crab in Creamy
Sauce

預備時間：**20 分鐘** ■烹調時間：**15 分鐘** ■份量：**4 人**
■ Preparation time：**20 mins**
■ Cooking time：**15 mins**
■ Serves：**4**

材料：

花蟹1隻（約12兩）
大伊麵1個
熱雞湯½杯
靈芝菇1盒
洋葱¼個
蒜蓉1茶匙
牛油2湯匙
花奶½杯

芡汁：

清雞湯1杯
生粉1湯匙
胡椒粉少許

Ingredients:

1 large crab (appox. 450g)
1 large e-fu noodle
½ cup hot chicken broth
1 packet lingzhi mushrooms
¼ onion
1 tsp minced garlic
2 tbsp butter
½ cup evaporated milk

Thickening Sauce:

1 cup chicken broth
1 tbsp cornstarch
A pinch of pepper

◥TIPS 貼士：

一定要食活蟹，因為死蟹會釋放出毒素，對身體無益，所以買回來的蟹應盡早將它殺掉，清洗好放冰箱保鮮。在家中多數用煎蟹方法，所以緊記蟹要吸乾水及灑少許生粉吸收多餘水份，再用高火煎封蟹件，然後加喜愛汁料同煮。

Only live crabs can be used for cooking as the dead ones will release toxic substances which are hazardous to health. Therefore, you must kill them shortly from the market, wash them thoroughly and then store them in the freezer. If you want to cook the crabs by pan-frying, you must pat them dry, sprinkle with cornstarch to absorb the excess moisture, fry them on high heat and then stir in your favourite sauce to cook.

做法：

1. 洋葱切幼粒；靈芝菇剪去菇腳，沖水。
2. 花蟹洗淨斬件，吸乾水，灑少許鹽、胡椒粉調味，放冰箱內備用。
3. 靈芝菇放滾水內拖水，隔水備用；伊麵放滾水內拖軟，隔水，放窩內，注入熱雞湯浸至入味，上碟。
4. 蟹件炸前灑上生粉，放6湯匙滾油內半煎炸至熟透，盛起。
5. 熱溶斗油，下洋葱、蒜蓉及靈芝菇爆香，蟹件同煮，濳酒，加入芡汁煮滾，最後加入花奶拌勻材料不用濃稠，淋伊麵上趁熱品嘗。

Method:

1. Cut onion into fine dices; remove the stems of lingzhi mushrooms and then rinse with water.
2. Wash crab thoroughly, chop into pieces, pat dry, sprinkle with a little salt and pepper to taste, chill in the refrigerator and then set aside.
3. Blanch lingzhi mushrooms in boiling water, strain and then set aside; blanch e-fu noodles in boiling water until softened, strain, place it in a pot, fill in hot chicken broth and leave the noodles in the broth until tasty. Transfer the noodles to a dish.
4. Sprinkle the crab with some cornstarch. Heat up 6 tbsp of oil in a wok, put in crab to deep-fry until cooked. Take it out.
5. Melt butter in the wok, put in onion, minced garlic and lingzhi mushrooms, and stir-fry. Return the crab to the wok, splash in some wine, add in thickening sauce and bring it to a boil. Finally, add in evaporated milk and stir until the sauce thickens. Pour it over the noodles and serve hot.

金銀蒜蒸海中蝦

Steamed Shrimps with Garlics

預備時間：**20 分鐘** ■烹調時間：**7 分鐘** ■份量：**4 人**

■ Preparation time：**20 mins**
■ Cooking time：**7 mins**
■ Serves：**4**

材料：
海中蝦8兩 (300克)
粉絲1細扎
蒜蓉6湯匙
芫荽1棵
葱1條

調味：
鹽1茶匙
糖1/4茶匙
胡椒粉少許

上湯：
滾水1/2杯
雞粉1茶匙
胡椒粉、麻油少許

Ingredients:
300g medium-size shrimps
1 small bunch of vermicelli
6 tbsp minced garlic
1 coriander
1 sprig spring onion

Seasonings:
1 tsp salt
1/4 tsp sugar
A pinch of pepper

Broth:
1/2 cup boiling water
1 tsp chicken powder
A little pepper and sesame oil

◥TIPS 貼士：

蒸蝦和白灼蝦，要夠鮮味，所以最好選用游水或剛死去的海中蝦。蒜茸要先略爆以去除生蒜味，有部份還要炸至金黃色，便成金蒜，以增強香味，粉絲要先加味浸泡過，因蒸時間不長，如此才可使各材料味道混合得好。

If you want to cook the shrimps by boiling, you must use live medium sea shrimps or those which have died before long for a fresh taste. The minced garlic has to be fried slightly to remove its strong flavour and some of it has to be deep-fried until browned to make golden garlic for a stronger flavour. Since the cooking time is short, vermicelli has to be soaked in broth before cooking so that all ingredients can blend well with each other.

做法：

1. 芫荽切碎；葱切粒；蒜頭剁蓉。

2. 燒6湯匙油，加半份蒜蓉爆至金黃色，隔油備用；剩餘滾油與其餘蒜蓉加調味拌勻。

3. 粉絲先用水浸軟，隔水，再加滾上湯浸至入味，放大碟上。

4. 中蝦沖洗乾淨，修剪好蝦鬚及腳，蝦背切開邊，去黑腸，蝦肉塗滿調味蒜蓉。

5. 中蝦排放粉絲面，隔水大火蒸4分鐘，灑上芫荽、葱粒及炸香蒜茸，蓋好出煙1分鐘，淋少許滾油及豉油調味，趁熱享用。

Method.

1. Chop coriander finely, cut spring onion into dices; chop garlic finely.

2. Heat up 6 tbsp of oil, add in half portion of minced garlic and stir-fry until golden brown, strain excess oil and then set aside. Mix the remaining oil with the rest of the minced garlic and seasonings.

3. Soak vermicelli in water until softened, strain, soak in hot broth until tasty, and then transfer to a large dish.

4. Wash shrimps thoroughly, trim shrimps antennae and legs, cut their back open, remove intestines, and then smear with seasoned minced garlic.

5. Line the shrimps on vermicelli, steam above water on high heat for 4 mins, sprinkle with coriander, spring onion dices and deep-fried minced garlic, cover with a lid and steam for another 1 min. Sprinkle a little boiling oil and soy sauce to taste. Serve hot.

黃金蝦

Golden Shrimps

預備時間：**10分鐘**　烹調時間：**10分鐘**　份量：**4人**
Preparation time：**10 mins**
Cooking time：**10 mins**
Serves：**4**

材料：
基圍蝦8兩 (300克)
熟鹹蛋黃3隻
牛油1湯匙

調味：
鹽1/4茶匙
胡椒粉少許
生粉2湯匙 (後下)

Ingredients:
300g shrimps
3 cooked salted egg yolks
1 tbsp butter
Seasonings:
1/4 tsp salt
A pinch of pepper
2 tbsp cornstarch (to be added later)

⊠TIPS 貼士：

用游水基圍蝦除夠新鮮外，可不用修剪蝦腳，蝦殼又夠薄身，吃時幾乎也可連同惹味酥香的鹹蛋黃醬料一起吃掉。除了蒸，可嘗試用快速煮熟鹹蛋黃的方法，只要把鹹蛋黃切成小粒，入微波爐以高火煮1-2分鐘便成，回鑊時要加牛油兜炒，鹹蛋黃碎便會沾在蝦上，快捷方便。

Live fresh water shrimps have soft legs, so you don't need to trim its legs before cooking. Besides, their shells are soft and you can almost swallow them together with the tasty salted egg yolk sauce. Other than steaming, the salted egg yolk can be cooked by a quick and easy method, for instance, cut salted egg yolk into small dices, cook it in an microwave oven on high heat for 1 to 2 minutes and then stir-fry it with butter and shrimps in a pan.

做法：

1. 鹹蛋黃隔水蒸10分鐘至熟透，趁熱壓碎備用。
2. 基圍蝦沖洗乾淨，加調味拌勻。
3. 蝦吸乾水，灒上乾生粉，放1杯燒油內炸至金黃色，收出，待油翻滾，再回鑊炸至脆身，隔油。
4. 用鑊燒熱1/2湯匙油及牛油，加鹹蛋黃碎以中火推至起泡，蝦仔回鑊快手兜炒至乾身，便成佐酒住餚。

Method:

1. Steam salted egg yolk above water for 10 mins until cooked, smash it while hot and then set aside.
2. Wash shrimps thoroughly, add in seasonings and mix well.
3. Pat the shrimps dry, sprinkle with cornstarch, deep-fry them with a cup of boiling oil and then take them out. When the oil is boiling again, return the shrimps to wok to deep-fry until crispy, and then strain.
4. Heat up 1/2 tbsp of oil and butter, add in salted egg yolk crumbs to stir-fry on medium heat until bubbles appear. Return shrimps to the wok and stir-fry quickely until the sauce is absorbed. It can be served as an accompaniment of drinks.

鮑汁蜜豆炒鮮魷

Stir-fried Fresh Squid
with Honey Peas in
Abalone Sauce

預備時間：**20 分鐘** ■烹調時間：**15 分鐘** ■份量：**4 人**

■ Preparation time：**20 mins**

Cooking time：**15 mins**

■ Serves：**4**

材料：

鮮魷魚6兩 (240克)
蜜糖豆6兩 (240克)
紅蘿蔔花8片
粟米仔4條
薑2片
蒜頭1粒

醃料：

鹽1/4茶匙
薑汁酒1湯匙
胡椒粉少許

芡汁：

生粉1/4茶匙
水1/2湯匙
鮑魚汁2½湯匙

Ingredients:
240g fresh squid
240g honey peas
8 slices carrot
4 baby corns
2 slices ginger
1 garlic

Marinade Ingredients:
1/4 tsp salt
1 tbsp ginger wine
A pinch of pepper

Thickening Sauce:
1/4 tsp cornstarch
1/2 tbsp water
2½ tbsp abalone sauce

◤**TIPS 貼士：**

揀魷魚要淺色雪白才夠新鮮，已轉紫紅色的較差一點，切件魷魚要剬十字花，除美觀外，還令魷魚容易消化。經醃過的魷魚要先拖水或泡油，以鎖着水份，到回鑊炒時使不會大量出水，影響口感。

Select fresh squids in snowy white colour rather than the purplish ones. Slit crosses on its surface for a better presentation and ease of digestion. The marinated squids have to be blanched in boiling water or hot oil to block their moisture, otherwise, they will secrete lots of liquid during stir-frying and thus, damaging their texture.

做法：

1. 清洗魷魚內臟，撕去外衣，剬十字花，切件，拌入醃料醃片刻。

2. 蜜糖豆撕去筋；粟米仔斜切開半；蒜頭切片；薑及紅蘿蔔切花。

3. 燒半鑊滾水，加入鹽及雞粉各1/2茶匙，把蜜糖豆、紅蘿蔔、粟米仔等拖水，隔起。

4. 再燒滾水，下魷魚拖水至捲起，吸乾水。

5. 燒2湯匙油，下薑、蒜及所有材料回鑊爆炒，潷酒，加芡汁兜炒令乾身，便可上碟。

Method:

1. Clean the internal organs of squid, tear off its skin, slash crosses on its back, cut into pieces, mix it with marinade ingredients and then leave it for a while.

2. Tear off the hard strings of honey peas, cut baby corns diagonally in half; cut garlic into slices; cut ginger and carrot into floral patterns

3. Heat up half wok of water, add in 1/2 tsp of salt and chicken powder, add in honey peas, carrot, baby corns to blanch for a while, and then strain.

4. Heat up some water to blanch squid in it until the squid rolls up, and then pat dry.

5. Heat up 2 tbsp of oil, add in ginger, garlic and all the ingredients to the wok to stir-fry, splash in some wine, add in thickening sauce and stir-fry until the sauce reduces. Transfer to a dish and serve.

冬菜九肚魚粉絲湯

Preserved Cabbage and Bombay Duck Fish in Vermicelli Soup

預備時間：**10 分鐘** ■烹調時間：**15 分鐘** ■份量：**4 人**

■ Preparation time：**10 mins**

■ Cooking time：**15 mins**

■ Serves：**4**

材料：

九肚魚12兩 (450克)
豆腐泡2兩 (80克)
粉絲1扎
冬菜1湯匙
薑3片
唐芹 (芹菜)1條
滾水3杯

醃料：

鹽½茶匙
魚露1茶匙
胡椒粉少許

Ingredients:

450g Bombay duck fish
80g beancurd puff
1 bunch vermicelli
1 tbsp preserved cabbage
3 slices ginger
1 Chinese celery
3 cups boiling water

Marinade Ingredients:

½ tsp salt
1 tsp fish sauce
A pinch of pepper

◥TIPS 貼士:

平常煮魚湯前要先將魚煎香，方可有奶白色的效果，但是煮九肚魚卻不用了，因九肚魚肉質較軟腍，多水份，不可能煎香的，所以會用鹽油水方法去煮，意思是用油爆香薑絲及鹽再加少量水加材料大火煮，這樣可把海鮮的鮮味迫出來。嚴格來説，這個不是湯，是漁夫常用的煮海鮮方法，夠方便快捷。

Fish are usually pan-fried before cooking into a milky fish soup. However, this cannot be applied to Bombay duck fish due to their soft and watery flesh. Therefore, we have to stir-fry shredded ginger with salt in oil before adding water to boil on high heat, and then put in Bombay duck fish to cook so as to bring out the fresh taste of fish. This handy method is commonly used by fishermen.

做法：

1. 粉絲浸軟；九肚魚洗淨去除頭尾，切段，加醃料同醃10分鐘。

2. 薑切絲；芹菜切碎；豆腐泡切厚片；冬菜沖洗乾淨備用。

3. 燒4湯匙油，放下豆腐泡兜炒至脆身；盛起，用餘下油炒香薑絲，落¼茶匙鹽及滾水煮滾，然後加九肚魚以大火煮10分鐘至湯呈奶白色，加冬菜、粉絲煮至熟透，最後加入芹菜粒，以胡椒粉、魚露調味上碟，灑上脆炸豆腐泡趁熱享用。

Method:

1. Soak vermicelli to soften; wash Bombay duck thoroughly, remove its head and tail, cut into chunks, mix with marinade ingredients and then leave it for 10 mins.

2. Cut ginger into shreds; chop Chinese celery finely; cut beancurd puff into thick slices; wash preserved cabbage thoroughly and then set aside

3. Heat up 4 tbsp of oil, put in beancurd puffs to stir-fry until crispy, and then take it out. Stir-fry ginger shreds in the remaining oil, put it ¼ tsp of salt and boiling water, and bring it to a boil. Put in Bombay duck to cook on high heat for 10 mins until the soup turns milky white, add in preserved cabbage and vermicelli, to cook until cooked. Finally add in Chinese celery dices and pepper and fish sauce to taste. Transfer to a pot, sprinkle the crispy beancurd puffs on top and serve hot.

北菇三文魚頭煲

Stewed Salmon Head
with Mushrooms

↘

預備時間：**10 分鐘** ■烹調時間：**30 分鐘** ■份量：**4 人**

■ Preparation time：**10 mins**
■ Cooking time：**30 mins**
■ Serves：**4**

材料：
三文魚頭1個
冬菇8隻
芋絲1包
實豆腐1件
生菜8兩（300克）
薑4片
乾葱頭4粒
辣椒仔2隻

醃料：
生抽1湯匙
薑汁酒1湯匙
胡椒粉少許
生粉1湯匙

調味：
水1½杯、雞粉1茶匙
鹽¼茶匙、糖½茶匙
生抽1湯匙、胡椒粉少許

芡汁：
生粉1湯匙
水2湯匙

Ingredients:
1 salmon head
8 shiitake mushrooms
1 pack shirataki
1 pc hard tofu
300g lettuce
4 slices ginger
4 shallots
2 chilies

Marinade Ingredients:
1 tbsp soy sauce
1 tbsp ginger wine
A pinch of pepper
1 tbsp cornstarch

Seasonings:
1½ cups water
1 tsp chicken powder
¼ tsp salt
½ tsp sugar
1 tbsp soy sauce
A pinch of pepper

Thickening Sauce:
1 tbsp cornstarch
2 tbsp water

蛋、豆腐和湯羹

瓜菜

水產

肉類

甜品

◣TIPS 貼士：

三文魚頭在壽司店或大型超市購買的比凍肉舖所買到的品質好得多。它可用油炸的方法，或放入烤爐烤香再回鍋炆，效果也不錯。

Salmon heads from sushi stores or large supermarkets are of better quality than the ones from frozen food stores. Instead of pan-frying salmon head, you may also deep-fry or roast in an oven before stewing in a pot.

做法：

1. 三文魚頭開半，塗勻醃料待醃片刻。

2. 乾葱、辣椒仔略拍；冬菇浸軟去蒂；芋絲沖洗乾淨。

3. 豆腐切厚件，灑少許鹽同醃，吸乾水，放熱油內煎香，上碟。

4. 燒肉燒熱油，下薑片及魚頭煎香，加乾葱及冬菇略燦，注入調味料並好炆10分鐘，加入辣椒、芋絲、豆腐再煮至入味，備用。

5. 用砂鍋燒少許油，放下生菜片，倒入魚頭等材料蓋好煮至菜軟身，埋芡汁，原鍋趁熱上桌。

Method:

1. Cut salmon fish head in half, smear it with marinade evenly and leave it for a while.

2. Slightly smash shallot and chilies; soak shiitake mushrooms to soften and then remove their stems; rinse shirataki with water.

3. Cut tofu into thick slices, sprinkle with a pinch of salt to marinate, pat dry, pan-fry it in hot oil until golden, and then transfer to a dish.

4. Heat up some oil in a wok, put in ginger slices and fish head to pan-fry until golden, and then put in shallots and shiitake mushrooms to stir-fry for a while. Add in seasonings, cover the wok with a lid to stew for 10 mins. Add in chilies, shirataki and tofu to cook until tasty, and then set aside.

5. Heat up a little oil in a clay pot, put in lettuce, and then pour in the fish head and accompaniments. Cover the pot to cook until the lettuce becomes soft, add in thickening sauce and bring it to a boil. Serve hot with the pot

紅燒划水
Braised Grass Carp

預備時間：**5 分鐘** ■烹調時間：**20 分鐘** ■份量：**4 人**
■ Preparation time：**5 mins**
■ Cooking time：**20 mins**
■ Serves：**4**

材料：	Ingredients:
鯇魚尾1條（約12兩）	1 grass carp (tail section, appox. 450g)
八角1粒	1 star aniseed
薑4片	4 slices ginger
葱4條	4 sprigs spring onions
芫荽1棵	1 coriander
調味料（A）：	Seasonings (A):
水¾杯	¾ cup water
生抽1湯匙	1 tbsp soy sauce
1老抽湯匙	1 tbsp dark soy sauce
糖1湯匙	1 tbsp sugar
調味料（B）：	Seasonings (B):
鎮江香醋1茶匙	1 tsp dark vinegar
麻油1茶匙	1 tsp sesame oil
芡汁：	Thickening Sauce:
生粉½茶匙	½ tsp cornstarch
水1湯匙	1 tbsp water

◤ TIPS 貼士：

鯇魚尾通常用來煮湯，蒸煮則多用魚腩，有道上海菜叫紅燒划水，划水意思是魚尾，只是用薑葱豉油炆煮便很香濃美味。在家中烹煮，因怕魚尾多骨，為了遷就小朋友，轉用魚腩也無妨。

Grass Carp tails are usually used for making soup while fish bellies are used for cooking. This Braised Grass Carp is a strong-flavoured Shanghaiese dish which braises grass carp fish tail with ginger, spring onion and soy sauce. However, as fish tail contains lots of bones, you may replace it with fish belly for children.

做法：

1. 葱切段；八角沖水分開成小粒。
2. 鯇魚尾洗淨抹乾，抹上少許鹽、胡椒粉及生粉調味。
3. 鑊內燒熱2湯匙油，放下魚尾以中火煎香兩面至甘香。
4. 八角、薑片及葱段放魚尾邊加少許油爆香，潷酒，注入調味料（A）蓋好以慢火煮8分鐘及至熟透，最後埋芡汁及拌入芫荽、調味料（B）以添美味，上碟，趁熱享用。

Method:

1. Cut spring onions into sections, rinse star aniseeds with water and then break into pieces.
2. Wash grass carp tail thoroughly, pat dry, smear with a little salt, pepper and cornstarch.
3. Heat up 2 tbsp of oil in a wok, put in fish tail to pan-fry on medium heat until both sides become golden.
4. Put star aniseeds, ginger slices and spring onions around the tail, add in a little oil to stir fry. Splash in some wine, add in seasoning (A), cover the wok with a lid and then cook on low heat for 8 mins until cooked. Finally, add in thickening sauce, coriander, seasonings (B) to taste. Transfer to a dish and serve hot.

甜酸松子黃魚

Sweet and Sour Yellow Croaker with **Pine Nuts**

預備時間：**20 分鐘**　烹調時間：**15 分鐘**　份量：**4 人**

☐ Preparation time：**20 mins**
☐ Cooking time：**15 mins**
☐ Serves：**4**

材料：

黃花魚1條 (約10兩重)
菠蘿片1片
青椒絲3湯匙
酸子薑3湯匙
酸蕎頭3粒
蒜蓉1茶匙
松子仁2湯匙
雞蛋1隻 (拂勻)
生粉½杯

醃料：

鹽¼茶匙
酒½湯匙
胡椒粉少許

甜酸汁：

水½杯
白醋1½湯匙
糖2湯匙
茄汁2湯匙
鹽⅛茶匙
生粉1茶匙

Ingredients:

1 yellow croaker (appox. 400g)
1 slice pineapple
3 tbsp shredded green pepper
3 tbsp preserved baby ginger
3 preserved bulbous onions
1 tsp minced garlic
2 tbsp pine nuts
1 egg (whisked)
½ cup cornstarch

Marinade Ingredients:

¼ tsp salt
½ tbsp wine
A pinch of pepper

Sweet and Sour Sauce:

½ cup water
1½ tbsp white vinegar
2 tbsp sugar
2 tbsp ketchup
⅛ tsp salt
1 tsp cornstarch

TIPS 貼士:

上海飯店的 "松鼠黃魚" 要按時價計，今時在市場也可輕易買到黃花魚，所以在家中自己做經濟得多。黃花魚要起骨並不難，只要在魚肚處用刀由頭部沿骨位向下切開，改用剪刀把骨剪下來便可。

In Shanghaiese restaurants, the price of this dish is seasonal and is quite expensive. Nowadays, as yellow croakers are widely available in the market, it is much more economical to make it at home. To bone the fish, simply slit its belly along its spine from head to tail with a knife and then remove the bones with a pair of scissors.

做法：

1. 菠蘿切粒；青椒、酸子薑及蕎頭切條。

2. 黃花魚洗淨，用尖刀從魚肚起出魚骨，魚身塗勻醃料，先沾上生粉，再塗勻蛋汁，最後再上滿生粉。

3. 燒1杯滾油，放下黃花魚半煎炸，期間用殼淋油魚頭魚尾，直至全條魚炸全香脆，隔油，上碟。

4. 燒少許油，下蒜蓉及青椒等切料煮，加入芡汁煮至濃稠，淋炸魚上，灑上松子仁，趁熱品嚐。

Method:

1. Cut pineapple into dices; cut green pepper, preserved baby ginger and preserved bulbous onions into strips.

2. Wash yellow croaker thoroughly, bone the fish from its belly with a pointed knife, smear it with marinade evenly, coat with cornstarch, egg liquid and then cover it with another layer of cornstarch again.

3. Heat up 1 cup of oil, put in yellow croaker to shallow fry, pour hot oil over the fish head and tail with a ladle frequently until the whole fish becomes crispy, strain out excess oil and then transfer the fish to a dish.

4. Heat up a little oil, put in minced garlic and green pepper to stir-fry, add in thickening sauce to cook until it thickens. Pour the sauce over the fish, sprinkle with pine nuts and serve hot.

豉汁豆泡蒸魚雲

Steamed Fish Head
with Beancurd Puff in
Black Bean Sauce

預備時間：**10分鐘** ■烹調時間：**15分鐘** ■份量：**4人**
- Preparation time：**10 mins**
- Cooking time：**15 mins**
- Serves：**4**

材料：
大魚頭1個（約12兩）
豆腐泡3兩（120克）
葱1條
芫荽1棵

醃料：
鹽1/3茶匙
糖1/4茶匙
生抽1茶匙
生粉2茶匙
胡椒粉、麻油少許

醬料：
紅椒絲1茶匙
蒜蓉2湯匙
豆豉1½湯匙
滾油2湯匙
蠔油1茶匙

Ingredients:
1 bighead carp head (appox. 450g)
120g beancurd puffs
1 sprig spring onion
1 coriander

Marinade Ingredients:
⅓ tsp salt
¼ tsp sugar
1 tsp soy sauce
2 tsp cornstarch
A little pepper and sesame oil

Sauce Ingredients:
1 tsp shredded red chilli
2 tbsp minced garlic
1½ tbsp preserved black beans
2 tbsp hot oil
1 tsp oyster sauce

◥TIPS 貼士:

一條大魚最有價值是個大魚頭，所以有魚販會一早收集魚頭再交給其他魚販出售，魚身一整條會很平價賣。但我喜歡買剛剛劏的，一定夠新鮮。蒸魚頭用豆腐泡墊底可吸收葱味的魚汁，令這道菜更加豐富，除了用豆腐泡，也可選擇芋絲或鮮腐竹。

As the most valuable part of a bighead carp is its head, some fish vendors will sell the fish heads to other vendors while selling the fish body at a bargain. However, I love using the head cut from a live fish for its freshness. This fish head should be steamed on a plate lined with beancurd puffs, which will absorb the tasty fish juice secreted in the process. You may replace beancurd puffs with shirataki or beancurd sheets as you like.

做法：

1. 芫荽、葱切碎；豆腐泡沖水備用。
2. 紅椒切絲，蒜頭剁蓉，豆豉沖水後壓爛，加滾油拌勻，再加蠔油調味備用。
3. 魚頭斬大件，洗淨，吸乾水，加醃料同醃，再加醬料拌勻。
4. 豆腐泡放碟內，魚頭放面，隔水大火蒸10分鐘及至熟透，灑上芫荽、葱蓋好再焗1分鐘，灑少許胡椒粉及生抽調味，趁熱享用。

Method:

1. Chop coriander and spring onions finely; rinse beancurd puffs with hot water and then set aside.
2. Shred red chili, chop garlic into mince, rinse preserved black beans and then smash into paste, add in hot oil and stir thoroughly, add in oyster sauce and then set aside.
3. Chop fish head into large pieces, wash thoroughly, pat dry, add in marinade ingredients and sauce ingredients, and mix well.
4. Put beancurd puffs in a dish with fish head on top. Steam it above water on high heat for 10 mins until cooked. Sprinkle with coriander and spring onions, and then cover it with a lid to steam for another 1 min. Sprinkle with a pinch of pepper and drizzle some soy sauce to taste. Serve hot.

POULTRY

冬菇魚肚蒸滑雞

Steamed Chicken
with Mushrooms
and Fish Maw

預備時間：**25分鐘** ■ 烹調時間：**15分鐘** ■ 份量：**4人**

■ Preparation time：**25 mins**
■ Cooking time：**15 mins**
■ Serves：**4**

材料：
光雞半隻 (約12兩)
冬菇4隻 (浸軟)
沙爆魚肚¾兩 (30克)
葱3條
薑6片

醃料：
鹽½茶匙
雞粉½茶匙
糖¼茶匙
生抽1茶匙
薑汁酒1湯匙
生粉1湯匙
胡椒粉、麻油少許

煨料：
水1½杯
雞粉2茶匙
酒1湯匙
胡椒粉少許

Ingredients:
One half of a chicken (appox. 450g)
4 shiitake mushrooms (soaked)
30g fried fish maw
3 sprigs spring onions
6 slices ginger

Marinade Ingredients:
½ tsp salt
½ tsp chicken powder
¼ tsp sugar
1 tsp soy sauce
1 tbsp ginger wine
1 tbsp cornstarch
A little pepper and sesame oil

Braising Sauce:
1½ cups water
2 tsp chicken powder
1 tbsp wine
A pinch of pepper

TIPS 貼士：

魚肚要選厚又重身的沙爆魚肚，浸水時要用重物壓着使容易吸收水才會浸軟。去除腥味時要用薑、葱、酒煮過，因為不宜煮太長時間，最好先把煨料煮至出味，才加入魚肚煮及浸至入味，如此可保持魚肚有軟綿綿口感。

Select fried fish maw which is thick and heavy. Soak it in water with a weight on it to facilitate water absorption and softening. Cook it with ginger, spring onion, wine and seasonings to remove its fishy smell. As the fish maw will become a mash if overcooked, we must cook the braising ingredients thoroughly before putting in the fish maw to cook briefly, and then leave it to absorb the tasty sauce.

做法：

1. 魚肚浸軟，搾乾水，切成大件。
2. 預備2條葱及4片薑放煨料內先煲5分鐘至出味，加魚肚滾起，熄火，蓋好焗至水涼，隔起。
3. 將餘下薑切絲；葱切度；冬菇切條。
4. 雞洗淨斬細件，加醃料同醃15分鐘，然後再與薑絲、冬菇拌勻；魚肚排放淺碟上，雞件等排放魚肚面，隔水大火蒸12分鐘及至熟透，葱度放面再蒸1分鐘，趁熱品嚐。

Method:

1. Soak fish maw until soften, squeeze out its water and then cut into large pieces.
2. Put 2 sprigs of spring onions and 4 slices of ginger into the braising sauce to boil for 5 mins until their flavour comes out. Add in fish maw, bring it to a boil, turn off the heat, cover with a lid and then leave it to cool. Take out fish maw.
3. Cut the remaining ginger into shreds, spring onions into sections and shiitake mushrooms into strips.
4. Wash chicken thoroughly, cut it into small pieces, add in marinade and leave it for 15 mins. Then mix it with shredded ginger and shiitake mushrooms. Line fish maw orderly on a plate with chicken placed on top evenly. Steam the dish above water on high heat for 12 mins until cooked. Sprinkle spring onions on top and steam for another 1 min. Serve hot.

西汁雞扒

Pan-fried Chicken Fillet
with Sweet and Sour
Sauce

預備時間：**15分鐘** ■烹調時間：**15分鐘** ■份量：**4人**

■ Preparation time：**15 mins**
■ Cooking time：**15 mins**
■ Serves：**4**

材料：

雞扒12兩（450克）
細洋葱1個
蒜蓉2茶匙匙
生粉2湯匙

醃料：

鹽1/4茶匙
美極鮮醬油1湯匙
黑胡椒碎1/4茶匙
生粉1湯匙

芡汁：

水1/3杯
鹽1/4茶匙
糖1湯匙
茄汁2湯匙
OK汁1湯匙
喼汁1/2湯匙

Ingredients:
450g chicken fillets
1 small onion
2 tsp minced garlic
2 tbsp cornstarch

Marinade Ingredients:
1/4 tsp salt
1 tbsp Maggi seasoning sauce
1/4 tsp grounded black pepper
1 tbsp cornstarch

Thickening Sauce:
1/3 cup water
1/4 tsp salt
1 tbsp sugar
2 tbsp ketchup
1 tbsp OK sauce
1/2 tbsp Worcestershire sauce

◣ TIPS 貼士：

雞扒即雞髀肉。這道菜最好要保留雞皮，如果怕肥膩可選用雞下髀肉，因為雞下髀肉較瘦，口感較爽口，煎前再加生粉令雞皮更脆口。

Chicken thigh meat is used to make a fillet. The skin should be kept intact for a better taste. You may select the lower thigh meat for its lean and elastic texture. Coat the meat with some cornstarch before pan-frying can make the chicken skin more crispy.

做法：

1. 洋葱切絲；雞扒切成大件，加醃料拌勻。

2. 雞扒臨煎前再加生粉拌勻。

3. 燒熱鑊下3湯匙油，雞扒落鑊煎至兩面金黃色及至九成熟，盛起，用餘下油爆香洋葱絲及蒜蓉，雞扒回鑊，濺酒，加芡汁兜至香濃入味，便可上碟。

Method:

1. Shred onion. Cut chicken fillets into large pieces and mix with marinade ingredients.

2. Coat chicken fillets with cornstarch before pan-frying.

3. Put 3 tbsp of oil into a hot pan, add in chicken fillets to pan fry until both side are brown and are almost cooked, and then take them out, then stir fry the onion and minced garlic with the remaining oil, return chicken fillets to cook, splash in some wine, add in thickening sauce and stir-fry until aromatic and tasty. Transfer to a dish and serve.

沙薑鹽水雞

Salted Chicken with Galangal

預備時間：**30 分鐘** ■烹調時間：**25 分鐘** ■份量：**6 人**
■ Preparation time：**30 mins**
■ Cooking time：**25 mins**
■ Serves：**6**

材料：

冰鮮雞1隻
粗鹽10兩（400克）
糖2湯匙
沙薑片4片
八角2粒
清水20杯
甘筍絲1杯
生菜絲1杯

醃料：

紹興酒1湯匙
薑汁1湯匙

沙薑油：

熟油3湯匙
沙薑粉2湯匙
雞粉1/8茶匙
鹽1/8茶匙

Ingredients:

1 chilled chicken
400g coarse salt
2 tbsp sugar
4 slices galangal
2 star aniseed
20 cups water
1 cup shredded carrot
1 cup shredded lettuce

Marinade Ingredients:

1 tbsp Shaoxing wine
1 tbsp ginger juice

Galangal Oil:

3 tbsp cooked oil
2 tbsp galangal powder
1/8 tsp chicken powder
1/8 tsp salt

TIPS 貼士：

現今潮流漸漸轉食冰鮮雞，如果烹調恰當，味道也不會比鮮雞差。正如這個鹽水雞，因有大量鹽份把雞的雪味辟除，浸雞時將雞浸大滾水內煮片刻，再把雞取起，待水滾後再把雞浸入，稱為"灌水"，如此做三次，目的是鎖住肉汁，然後用慢火浸熟，可用竹籤穿過雞脾，如滲出的肉汁帶有紅色，表示未夠熟透，需要加時。

If being processed properly, chilled chickens can taste as great as the fresh ones. In the making of this salted chicken, the large amount of salt helps to remove the cold meat smell of the chicken. After that, the chicken is poached in boiling water briefly and then taken out; when the water is boiling again, repeat the poaching process two more times so as to block the meat juice of chicken. Then simmer the chicken in the liquid on low heat until cooked. If red meat juice seeps out when a bamboo skewer is pierced through the chicken thigh, it means that the chicken is not fully cooked and has to be simmered for a while longer

做法：

1. 光雞洗淨抹乾，用醃料塗勻雞身內外，醃15分鐘。

2. 用大鍋燒滾水，加入沙薑片、八角、鹽及糖先煮10分鐘至出味，將雞放鹽水內灌水三次，待水翻滾後改用慢火浸18-20分鐘至全熟透，取出，待凍透後便可拆肉絲及斬件上碟。

3. 預備生菜絲及甘筍絲伴雞肉邊，可作涼菜享用，食時蘸沙薑油增添美味。

Method:

1. Wash chicken thoroughly, pat dry and then smear its exterior with marinade to marinate for 15 mins.

2. Heat up water in a large pot, add in galangal slices, star aniseed, salt and sugar to cook for 10 mins. Put in chicken and fill it with the salted liquid three times. When the water is boiling, cook the chicken on low heat for 18-20 mins until cooked. Take the chicken out, leave to cool. Then shred its meat and cut into pieces before transferring to a dish.

3. Surround the chicken with shredded lettuce and carrot as accompaniments. Serve it with galangal oil dipping sauce for a more exotic taste.

玫瑰豉油雞

Chicken with Soy
Sauce and Rose Essence
Wine

預備時間：**15 分鐘** ▉烹調時間：**20 分鐘** ▉份量：**4 人**
▉ Preparation time：**15 mins**
▉ Cooking time：**20 mins**
▉ Serves：**4**

材料：
光雞半隻 (約1斤)
薑3片
葱2條
八角2粒
葱絲適量

豉油料：
水1/2杯
生抽1/2杯
老抽1/4杯
玫瑰露酒2湯匙
冰糖2兩 (80克)

Ingredients:
Half of a chicken (appox. 600g)
3 slices ginger
2 sprigs spring onion
2 star aniseed
Some shredded spring onions

Soy Sauce Ingredients:
1/2 cup water
1/2 cup soy sauce
1/4 cup dark soy sauce
2 tbsp rose essence wine
80g rock sugar

TIPS 貼士：

烹煮半隻雞比煮一隻雞方便快捷得多，可是雞肉熟後會略為收縮，適合人少家庭享用，做時要注意先預備豉油雞汁至出味才把雞拖水，雞要趁熱放入豉油內上色，想保証完全熟透也可熄火後多浸5分鐘。如果要煮一隻雞便要多加汁料，還要多煮10分鐘。

Cooking half a chicken is a lot faster than a whole one. However, as the meat will contract when cooked, it is more suitable for a small household. As the chicken has to be coloured with soy sauce when it is hot, you have to get the chicken sauce ready before blanching the chicken in boiling water. You may poach the chicken in hot water for 5 more minutes after turning off the heat to ensure the chicken is well done. If you want to make a whole chicken, you have to cook it for 10 minutes longer than the time specified.

做法：

1. 光雞洗淨，放半鑊滾水內拖水，取出。
2. 用鑊燒少許油，卜薑、葱、八角爆香，注入豉油料煮滾，以慢火先煮3分鐘至汁香濃，加入雞淋勻豉油上色，雞皮向下，蓋好以慢火先煮5分鐘，反轉另一面煮5分鐘，再反轉煮5分鐘至熟透。
3. 取出雞，略凍後便可斬件，放碟上，用葱絲作裝飾，澆上少許豉油汁及麻油品嚐。

Method:

1. Wash chicken thoroughly, blanch it in a wok half-filled with boiling water for a while and then take it out.
2. Heat up a little oil in a wok, add in ginger, spring onion and star aniseed to stir-fry. Pour in soy sauce ingredients and bring it to a boil. Simmer on low heat for 3 mins until the sauce condenses. Then add in chicken and pour the sauce over it to colour it. Turn the chicken with skin downwards, cover the wok with its lid and simmer on low heat for 5 mins. Turn the chicken over and cook for another 5 mins. Repeat this turning and cooking process until the chicken is cooked.
3. Take the chicken out, leave to cool, cut into pieces and then transfer to a plate. Garnish with shredded spring onions, and then drizzle some soy sauce and sesame oil to taste.

青芥末雞柳炒露筍

Stir-fried Chicken and
Asparagus with Wasabi

預備時間：**15分鐘** 烹調時間：**10分鐘** 份量：**4人**

■ Preparation time：**15 mins**
■ Cooking time：**10 mins**
■ Serves：**4**

材料：
雞柳4兩 (150克)
鮮露筍6兩 (240克)
薑2片
葱1條
日本青芥末2湯匙

醃料：
鹽¼茶匙
糖¼茶匙
生抽1茶匙
薑汁酒2茶匙
生粉1茶匙
胡椒粉及麻油少許

調味：
滾水¼杯
糖½茶匙
雞粉½茶匙

Ingredients:
150g chicken fillets
240g fresh asparagus
2 slices ginger
1 sprig spring onion
2 tbsp wasabi

Mariande Ingredients:
¼ tsp salt
¼ tsp sugar
1 tsp soy sauce
2 tsp ginger wine
1 tsp cornstarch
A little pepper and sesame oil

Seasonings:
¼ cup boiling water
½ tsp sugar
½ tsp chicken powder

■TIPS 貼士：

鮮蘆筍要選美國粗大的，但不是經常有貨，我們亦可用粗芥蘭修淨外皮切片炒，一樣爽脆。鮮蘆筍味清淡，要加調味先煮才回鑊兜炒，加入日本青芥末是個很美妙的配搭。

It is best to use American asparagus which is big and thick. However, it is not always available for sale. Instead we can also replace it with thick kale, which is skinned and sliced, for an equally crunchy texture. Since fresh asparagus is light in taste, it has to be cooked with seasonings before stir-frying in a wok. Meanwhile, it goes perfectly well with wasabi.

做法：

1. 雞柳切條，加醃料拌勻。
2. 薑切花；葱切段；鮮露筍修去硬皮，斜切厚片。
3. 燒1湯匙油，加薑花及露筍爆香，加調味料蓋好以大火煮1分鐘，盛起，隔去水份。
4. 燒2湯匙油，下雞柳爆至八成熟，葱段、露筍回鑊，濽酒，加入青芥末快手兜炒至入味，便可上碟。

Method:

1. Cut chicken fillets into strips and then mix with marinade.
2. Cut ginger into floral shapes and spring onions into sections. Peel the hard skin off the fresh asparagus and then cut into thick slices.
3. Heat up 1 tbsp of oil, add in ginger and asparagus to stir-fry, add in seasonings and cover it with a lid to cook on high heat for 1 min. Take it out and filter out the liquid.
4. Heat up 2 tbsp of oil, add in chicken fillets to stir fry until almost done, add in spring onion and asparagus, splash in some wine, add in wasabi and stir-fry quickly. Then transfer to a dish to serve.

原個南瓜咖喱雞盅

Curry Chicken
in Pumpkin

預備時間：**15 分鐘** ■烹調時間：**30 分鐘** ■份量：**4 人**

■ Preparation time：**15 mins**
■ Cooking time：**30 mins**
■ Serves：**4**

材料：

小南瓜1個 (約1½斤重)

雞髀肉8兩 (300克)

乾葱蓉1湯匙

咖喱醬1湯匙

洋葱½個

青椒½隻

紅甜椒½隻

蘑菇1罐

水¾杯

椰漿½杯

花奶¼杯

醃料：

鹽⅓茶匙、糖¼茶匙

生抽1茶匙、生粉½湯匙

薑汁酒½湯匙、胡椒粉少許

調味：

雞粉1茶匙

鹽½茶匙

糖½茶匙

Ingredients:

1 small pumpkin (appox. 800g)

300g chicken thigh meat

1 tbsp minced shallot

1 tbsp curry paste

½ onion

½ green pepper

½ red bell pepper

1 can mushrooms

¾ cup water

½ cup coconut milk

¼ cup evaporated milk

Mariande Ingredients:

⅓ tsp salt

¼ tsp sugar

1 tsp soy sauce

½ tbsp cornstarch

½ tbsp ginger wine

A pinch of pepper

Seasonings:

1 tsp chicken powder

½ tsp salt

½ tsp sugar

TIPS 貼士：

做這個菜用金黃色或綠色圓形小南瓜均可，用牙籤穿過瓜殼測試脸度，上桌時可用錫紙包裹以保護瓜殼完整，因小南瓜個子不大，餡料又要跟南瓜肉同吃，餡料要分兩次填入瓜盅內享用。

You may use small round pumpkins in golden or green colour. If a toothpick can pierce through the fruit, it is soft enough for cooking. Serve the whole fruit in aluminum foil to preserve its shape. As the pumpkin is small in size and its flesh can be eaten together with the filling, the filling has to be stuffed into the fruit in two separate servings.

做法：

1. 小南瓜切去¼份頂部，挖去瓜瓤，原個隔水蒸15分鐘至脸身，備用。

2. 雞肉切件，加醃料拌勻。

3. 洋葱、青、紅椒切塊；磨菇開半。

4. 燒1湯匙油，下乾葱蓉、洋葱及咖喱醬爆香，加入雞件炒透，再加其餘配料炒勻，注入水及調味料煮滾，改慢火炆10分鐘至入味，加入椰漿及花奶慢火煮片刻。

5. 把咖喱雞倒入南瓜盅內，可入焗爐或烤爐烤至面呈金黃色，食時，將軟脸南瓜肉挖出同吃，香甜美味。

Method:

1. Cut off quarter top portion of the pumpkin, scrape out the seeds and membrane inside, steam the whole pumpkin above water for 15 mins until softened and then set aside.

2. Cut chicken meat into pieces and mix with marinade.

3. Cut onion, greed and red peppers into pieces; cut button mushrooms in half.

4. Heat up 1 tbsp of oil, add in minced shallots, onion and curry paste to stir-fry, add in chicken to stir-fry thoroughly and add in the remaining ingredients to stir-fry. Pour in water and seasonings, and bring it to a boil. Simmer it on low heat for 10 mins, add in coconut milk and evaporated milk to cook on low heat for a short while.

5. Pour curry chicken into pumpkin, bake or grill it until the surface of pumpkin is browned. Serve curry chicken with the tender pumpkin flesh for a sweet delightful taste.

檸蜜雞翼

Chicken Wings with Honey and Lemon

預備時間：**30 分鐘** ■烹調時間：**15 分鐘** ■份量：**4人**

Preparation time：**30 mins**

Cooking time：**15 mins**

Serves：**4**

材料：
雞中翼10隻
蜜糖3湯匙
檸檬1個
炒香芝麻2茶匙
生粉2湯匙

醃料：
乾葱蓉2茶匙
鹽1茶匙
糖½茶匙
五香粉¼茶匙
生抽1½湯匙
生粉1湯匙
胡椒粉少許

Ingredients:
10 chicken wings (middle portion)
3 tbsp honey
1 lemon
2 tsp roasted sesame seed
2 tbsp cornstarch

Mariande Ingredients:
2 tsp minced shallot
1 tsp salt
½ tsp sugar
¼ tsp five-spice powder
1 ½ tbsp soy sauce
1 tbsp cornstarch
A pinch of pepper

TIPS 貼士：

如果怕炸雞翼用太多油，可改用平底鑊煎香，雞皮向下先煎，當排放好雞翼後可加蓋焗煎片刻，讓雞翼焗至半熟，反轉另一面繼續加少許油煎至香脆熟透。

If you find deep-frying too greasy, you may pan-fry the chicken wings with a little oil: first, line the wings on the pan with the skin facing down, fry with the lid on for a while until medium cooked; then turn the wings, add in a little oil to pan-fry until the whole chicken wings are fully cooked and crispy.

做法：

1. 雞翼洗淨，吸乾水，加入醃料同醃30分鐘。

2. 將半個檸檬切角；另外半個榨汁備用。

3. 雞翼加生粉拌勻，放半杯滾油內半煎炸至兩面金黃色及熟透，隔去油份。

4. 燒少許油，加入蜜糖及2茶匙檸檬汁煮至起泡，雞翼回鑊快手兜至乾身，上碟，灑上炒香芝麻及以檸檬作裝飾。

Method:

1. Wash chicken wings thoroughly, pat dry, mix with marinade and leave it for 30 mins.

2. Cut half of the lemon into wedges and squeeze out the juice from the other half. Set aside.

3. Mix chicken wings with cornstarch evenly, shallow-fry them with half cup of oil until both sides are browned. Strain out excess oil.

4. Heat up a little oil, add in honey and 2 tsp of lemon juice to cook until bubbles appear. Return chicken wings to the wok to stir-fry until the sauce condenses. Transfer to a plate, sprinkle some roasted sesame seed on top and garnish with lemon wedges.

脆芒鴕鳥肉
Stir-fried Ostrich Meat
with Mango

預備時間：**20 分鐘** ▓烹調時間：**10 分鐘** ▓份量：**4 人**

▓ Preparation time：**20 mins**

▓ Cooking time：**10 mins**

▓ Serves：**4**

材料：
鴕鳥肉4兩 (150克)
油條1條
青椒½隻
芒果2個
蒜頭1粒

調味：
生粉½湯匙
生抽2茶匙
糖¼茶匙
胡椒粉、麻油少許
油1茶匙 (後下)

醬汁：
泰國甜辣雞醬3湯匙
茄汁1湯匙

Ingredients:
150g ostrich meat
1 twisted doughnut
½ green pepper
2 mangoes
1 garlic

Seasonings:
½ tbsp cornstarch
2 tsp soy sauce
¼ tsp sugar
A little pepper and sesame oil
1 tsp oil (add in later)

Sauce Ingredients:
3 tbsp sweet chilli sauce for chicken
1 tbsp ketchup

TIPS 貼士：

鴕鳥肉是高蛋白質、低脂肪、低膽固醇的健康肉食，食法近似牛肉，多用作火鍋。它可在日式超市購買得到，可是價錢並不便宜，所以未能普及。

Ostrich meat is a healthy and nutritive food which is rich in protein but low in fat and cholesterol level. It tastes like beef and is mostly used in hot-pot. Although it is available in Japanese supermarkets, it is not very common due to its relatively high prices.

做法：

1. 鴕鳥肉吸乾水後橫紋切片，加醃料同醃15分鐘。
2. 芒果去皮切厚片，灑上1茶匙白醋拌勻；青椒切條；蒜頭切片；油條切厚片。
3. 燒熱5湯匙油，放下油條兜炒至香脆，取出隔油；用餘下油爆鴕鳥肉至八成熟，盛起隔油。
4. 燒1湯匙油，下蒜片、鴕鳥肉及青椒炒透，加入醬汁、芒果及油條兜勻，便可上碟。

Method:

1. Pat ostrich meat to dry, slice it against the grain, mix with marinade and leave for 15 mins.
2. Peel mangoes, cut into thick slices, and then mix with 1 tsp of white vinegar. Cut green pepper into strips and garlic into slices. Cut twisted doughnut into thick slices.
3. Heat up 5 tbsp of oil, add in twisted doughnut to stir fry until crispy, and then take it out. Stir fry ostrich meat with the remaining oil in the wok until almost done, take it out and strain.
4. Heat up 1 tbsp of oil, add in garlic slices, ostrich meat and green pepper to stir-fry thoroughly. Add in sauce ingredients, mangoes and twisted doughnut to stir-fry evenly. Transfer to a plate to serve.

辣汁乳鴿

Baby Pigeons
in Chili Sauce

預備時間：**10 分鐘** ■烹調時間：**50 分鐘** ■份量：**4 人**

■ Preparation time：**10 mins**

■ Cooking time：**50 mins**

■ Serves：**4**

材料：
乳鴿2隻
乾蔥頭6粒
辣椒仔2隻

浸鴿料：
水10杯
八角2粒
薑2片
蔥2條
鹽1湯匙
紹興酒1湯匙

芡汁：
水1杯
雞粉½茶匙
生抽2茶匙
豆瓣醬2茶匙
蜜糖2湯匙

芡汁：
生粉1茶匙
水1湯匙

Ingredients:
2 pigeons
6 shallots
2 chilies

Soaking Ingredients for Pigeons:
10 cups water
2 star aniseed
2 slices ginger
2 sprigs spring onion
1 tbsp salt
1 tbsp Shaoxing wine

Thickening Sauce:
1 cup water
½ tsp chicken powder
2 tsp soy sauce
2 tsp chili bean sauce
2 tbsp honey

Thickening Sauce:
1 tsp cornstarch
1 tbsp water

TIPS 貼士：

辣汁乳鴿與紅燒乳鴿食法不同，辣汁乳鴿經浸熟後會再走油，然後加入香濃的醬汁同炆，不要求外皮杳脆，所以如果無新鮮乳鴿也可以用急凍乳鴿代替。

Pigeons are cooked in boiling water before blanching in hot oil, and then stewed in strong sauce. You may use chilled pigeons rather than fresh ones.

做法：

1. 辣椒仔略拍扁；乾蔥開半；乳鴿洗淨備用。

2. 用鍋先煮浸鴿料5分鐘，放下乳鴿大火煮滾，改用慢火浸煮20分鐘，取出，趁熱把乳鴿塗勻老抽上色，盛起，用吸油紙抹乾鴿腔內汁液。

3. 燒1杯滾油，放下乳鴿淋油至鴿皮呈金黃色，盛起；再下乾蔥炸香，隔油。

4. 燒少許油，卜乾蔥、辣椒仔、乳鴿及芡汁煮滾，以慢火炆煮5分鐘至入味，取出斬件上碟；辣汁加生粉水煮成薄芡，淋乳鴿面，趁熱享用。

Method:

1. Lightly crush chilies; cut shallots in half, wash pigeons thoroughly and set aside.

2. Cook soaking ingredients for pigeons in a wok for 5 mins and then put in pigeons to cook on high heat. When it is boiling, simmer it on low heat for 20 mins and then take it out. Brush some dark soy sauce on the pigeons when they are still hot. Then pat dry the inner part of pigeons with absorbent paper.

3. Heat up 1 cup of oil, put in pigeons and pour it with hot oil with a laddle until their skin is browned. Take them out. Put in shallots to fry and then drain.

4. Heat up a little oil, add in shallots, chilies, pigeons and thickening sauce to stew on low heat for 5 mins. Take out the pigeons, chop into pieces and then transfer to a dish. Cook chili sauce and cornstarch liquid into a sauce, pour it over the pigeons and serve hot.

酸薑梅子鴨

Braised Duck with
Preserved Ginger and
Plums

預備時間：**10 分鐘** ■烹調時間：**70 分鐘** ■份量：**4 人**
■ Preparation time：**10 mins**
■ Cooking time：**70 mins**
■ Serves：**4**

材料：
光鴨半隻 (約1½斤重)
蒜蓉2湯匙
磨豉醬1¼湯匙
酸梅子5粒
酸子薑2兩 (80克)
片糖½磚
茄汁1湯匙
紹興酒1湯匙

芡汁：
生粉1茶匙
水2湯匙

Ingredients:
One half of a duck (appox. 800g)
2 tbsp minced garlic
1¼ tbsp crushed yellow bean sauce
5 preserved plums
80g preserved baby ginger
½ pc cane sugar
1 tbsp ketchup
1 tbsp Shaoxing wine

Thickening Sauce:
1 tsp cornstarch
2 tbsp water

TIPS 貼士：

要燉鴨好味，一定要將鴨皮走過油，去除多餘油份才去燉或炆，用燉方法，肉質較完整和鬆軟。在家中，用全鴨走油，當然困難及危險，若將半隻鴨再開半，逐件走油，用油可以減少，對初學者容易掌握得多。

The duck must be blanched in hot oil to remove excess oil before double-boiling or stewing. Double-boiling is recommended as it can maintain the tenderness of the meat. Since it is quite dangerous to blanch a whole duck in hot oil at home, it is easier for beginners to blanch it in half individually. What's more, it takes less oil to blanch half a duck at a time.

做法：

1. 光鴨洗淨，鴨身斬開半，放入大鑊滾水內拖水5分鐘，取出，趁熱將鴨皮塗勻老抽上色，待涼。

2. 鑊內燒半杯滾油，將鴨件抹乾，鴨皮向下，鴨件一件件地淋油至皮呈金黃色，取出，用水沖去油漬，備用。

3. 片糖剁碎；梅子壓爛去核。

4. 燒2湯匙油，下蒜蓉、磨豉醬爆香，加入其餘材料煮至片糖溶化，把鴨回鑊與醬料兜勻，放淺碟內，用錫紙蓋好，隔水大火蒸45分鐘，把醬汁撥走留用，鴨斬件上碟。

5. 鴨醬汁回鑊內埋芡汁煮滾，醬汁淋鴨面，可多加酸薑伴吃，增添食慾。

Method.

1. Wash duck thoroughly, cut it in half, blanch in a large wok of boiling water for 5 mins, and then take it out. Brush it with dark soy sauce to colour when it is hot. Leave to cool.

2. Heat up half cup of oil in a wok, put in pat-dried duck pieces with skin downward, pour hot oil over duck pieces individually until browned and then take them out. Rinse with water to remove the oil and then set aside.

3. Chop cane sugar finely; mash plums and remove the cores.

4. Heat up 2 tbsp of oil, put in minced garlic and crushed yellow bean sauce to stir-fry, add in the remaining ingredients to cook until cane sugar melts. Put in duck pieces and sauce ingredients to stir-fry and then transfer to a shallow dish. Cover the dish with an aluminium foil, steam above water on high heat for 45 mins, scoop out the sauce and set aside. Chop the duck into pieces and transfer to a dish.

5. Put the duck sauce into the wok and cook with thickening sauce until it boils. Pour it over the duck and serve with preserved baby ginger for good appetite.

【家畜】

LIVESTOCK

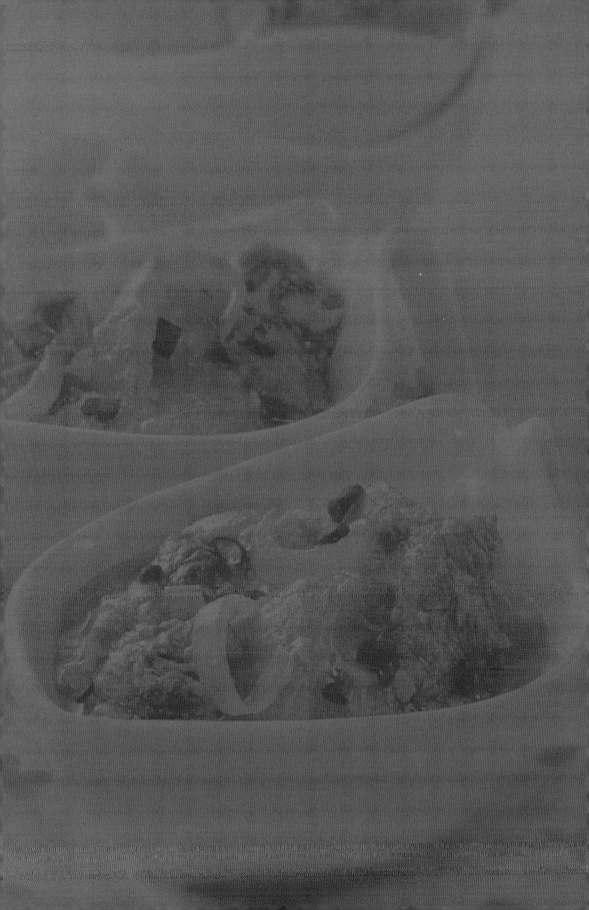

香煎馬蹄牛肉餅

Pan-fried Beef Cakes
with Water Chestnuts

預備時間：**15分鐘** ■烹調時間：**10分鐘** ■份量：**4人**
■ Preparation time：**15 mins**
■ Cooking time：**10 mins**
■ Serves：**4**

材料：

免治牛肉4兩 (150克)
馬蹄4粒
芫荽1棵
葱1條
新鮮麵包糠半杯
蛋汁3湯匙

醃料：

鹽1/4茶匙
糖1茶匙
黑胡椒碎1/4茶匙
美極鮮醬油1湯匙
喼汁1茶匙
生粉1湯匙
水2湯匙

Ingredients:

150g minced beef
4 water chestnuts
1 coriander
1 sprig spring onion
Half cup fresh bread crumbs
3 tbsp beaten egg

Marinade Ingredients:

1/4 tsp salt
1 tsp sugar
1/4 tsp ground black pepper
1 tbsp Maggi seasoning sauce
1 tsp worcestershire sauce
1 tbsp cornstarch
2 tbsp water

◥TIPS 貼士：

新鮮麵包糠是用去皮白方包磨碎而成，如果麵包太柔軟，可先烘乾後才容易磨碎，所以最好是用隔夜麵包。新鮮麵包糠加入牛肉餅內可減低用牛肉份量，也有吸收牛肉油份的效果，令牛肉餅更鬆軟。

Fresh breadcrumbs is made by grinding sliced white bread without crust. If the bread is too soft, you have to toast it before grinding. Therefore, stale bread is most suited. By mixing the beef with breadcrumbs, this can not only save the beef used but also absorb part of the beef fat, and thus enhancing the tenderness of the beef cakes.

做法：

1. 預備麵包去皮，切幼粒成包糠；芫荽、葱切碎；馬蹄去皮切粒。

2. 牛肉加醃料順一方向攪拌至黏性，再加麵包糠、蛋汁、馬蹄粒、芫荽及葱粒攪拌，分成十等份，搓成小球，再壓成餅形。

3. 用平底鑊燒熱油，下牛肉餅略煎，蓋好煎1分鐘，開蓋，反轉另一面再蓋1分鐘，開蓋，繼續煎至兩面金黃色及至熟透，上碟，食時可蘸茄汁或喼汁同吃。

Method:

1. Remove bread crust, cut bread into fine dices; chop coriander and spring onion finely; peel water chestnuts and then cut into dices.

2. Mix beef with marinade and stir unilaterally until sticky. Add in bread crumbs, beaten egg, water chestnuts, coriander and spring onion dices, mix thoroughly, divide into ten portions, shape into small balls and then press into a cake.

3. Heat up some oil in a pan, put in beef cakes to pan-fry slightly. Cover it with a lid to fry for 1 min, remove the lid, turn the cakes and then fry with the lid on for another min. Remove the lid and continue to fry until both sides are browned and cooked. Transfer to a dish and serve with ketcup or Worcestershire sauce.

香蒜玉豆牛仔骨

Veal Steak Ribs with
Garlic and Snap Beans

預備時間：**30 分鐘** 烹調時間：**15 分鐘** 份量：**4 人**
- Preparation time：**30 mins**
- Cooking time：**15 mins**
- Serves：**4**

材料：
牛仔骨8兩 (300克)
玉豆6兩 (240克)
三色甜椒絲3兩 (120克)
蒜蓉2湯匙
牛油1湯匙
拔蘭地酒少許

醃料：
生抽2茶匙
糖1/4茶匙
生粉2茶匙
黑胡椒碎1/2茶匙
油1/2湯匙 (後下)

調味料 (A)：
水1/2杯、雞粉1/2茶匙
鹽1/4茶匙、糖1/8茶匙

調味料 (B)：
水1湯匙、鹽1/8茶匙
雞粉1/4茶匙
麻油1/4茶匙

Ingredients:
300g veal steak ribs
240g snap beans
120g shredded red, green and yellow peppers
2 tbsp minced garlic
1 tbsp butter
A little brandy

Marinade Ingredients:
2 tsp soy sauce
1/4 tsp sugar
2 tsp cornstarch
1/2 tsp ground black pepper
1/2 tbsp oil (to be added later)

Seasonings (A):
1/2 cup water
1/2 tsp chicken powder
1/4 tsp salt
1/8 tsp sugar

Seasonings (B):
1 tbsp water
1/8 tsp salt
1/4 tsp chicken powder
1/4 tsp sesame oil

⬛TIPS 貼士:

來自美國的牛仔骨品質最好，每條牛仔骨都有三件骨，要挑選多肉而且分佈均勻的牛仔骨，否則只得兩件可食用。買預先包裝好的牛仔骨，不要買太大包，因內裏的品質會較差，最好在凍肉舖仔細挑選，品質較有保証。

US veal steak ribs are of the best quality with three pieces of bones in each rib. Select the ribs which are meaty and well-balanced or only two bones of a rib have meat on them. Don't buy a large packet of steak ribs as it usually contains poor quality ribs. If possible, try to select the ribs individually in frozen food stores to ensure only quality ones are used.

做法：

1. 牛仔骨用刀背拍鬆，切件，加入醃料同醃30分鐘。

2. 紅、黃、青、三色甜椒切絲；玉豆摘去頭尾筋，切段。

3. 燒1湯匙油，放下半湯匙蒜蓉及玉豆炒透，加入調味料(A)蓋好煮5分鐘，盛起，瀝乾水，上碟。

4. 燒熱鑊加1湯匙油，牛仔骨加油拌勻放鑊內以中火煎香至八成熟，盛起。

5. 以慢火熱溶牛油，加餘下蒜蓉及調味汁(B)煮滾，三色甜椒及牛仔骨回鑊，澆卜拔蘭地酒，以大火兜炒至乾身，放玉豆上，即成。

Method:

1. Tenderize the veal steak by pounding it with the back of a knife, cut into pieces, mix with marinade and leave for 30 mins.

2. Shred red, yellow and green peppers; remove the hard tissue of snap beans and then cut into sections..

3. Heat up 1 tbsp of oil, put in half tbsp of minced garlic and snap beans to stir-fry, add in seasonings (A), cover it with a lid to cook for 5 mins. Take it out, strain and then transfer to dish.

4. Add one more tbsp of oil to the heated wok, add in veal steak into the wok to pan-fry on medium heat until browned and almost cooked. Transfer to a dish.

5. Melt butter on low heat, add in the remaining minced garlic and seasonings (B). Add shredded peppers and veal steak, splash in some brandy and stir-fry on high heat until the sauce condenses. Put the veal steak on the dish of snap beans and serve.

清燉牛腩蘿蔔湯

**Double-boiled
Beef Briskets Soup**
with Turnips

預備時間：**10 分鐘** ■烹調時間：**2 小時** ■份量：**4 人**
- Preparation time：**10 mins**
- Cooking time：**2hrs**
- Serves：**4**

材料：

牛坑腩1斤 (600克)
白蘿蔔1斤 (600克)
薑6片
八角3粒
葱2條
冰糖1湯匙
紹興酒1湯匙
清雞湯4杯
芫荽1棵

Ingredients:

600g beef brisket
600g turnips
6 slices ginger
3 star aniseed
2 sprigs spring onions
1 tbsp rock sugar
1 tbsp Shaoxing wine
4 cups chicken broth
1 coriander

◣TIPS 貼士:

牛腩分有牛坑腩、枚頭腩、崩沙腩及車仔麵用的碎牛腩等，最上等要算是牛坑腩了。原件牛坑腩樣子整齊，半帶肉半帶筋，分佈均勻，味較香濃甘香，烹調牛腩應要把原件牛腩先煮熟後再切件加醬料烹煮，炆好的牛腩還要再焗透，便會更加鬆軟。

Select beef brisket with meat and tendons evenly distributed. Beef brisket should be cooked in whole piece before stewing in sauce in pieces; turn off the heat and leave the beef in the pot with a lid on for a longer while to tenderize the meat.

做法：

1. 牛腩洗淨，原件放滾水內拖水5分鐘，取出。

2. 用煲燒6杯滾水，加入3片薑、葱、八角、冰糖及紹興酒，放下牛腩慢火煮30分鐘，取出，切件，保留牛肉湯備用。

3. 牛腩件及其餘薑片放燉鍋內，注入燒滾清雞湯，蓋好先燉1小時。

4. 芫荽切碎；蘿蔔去皮切件，放牛肉湯內先煮5分鐘，隔起，加入牛腩湯內同燉至入味，下少許鹽、胡椒粉調味，灑上芫荽，原窩上桌。

Method:

1. Wash beef brisket thoroughly, blanch in boiling water for 5 mins and then take it out.

2. Heat up 6 cups of water in a pot and bring it to a boil. Add in 3 slices of ginger, spring onions, star aniseed, rock sugar and Shaoxing wine. Then add in beef brisket to stew on low heat for 30 mins, take it out and cut into pieces. Keep the beef stock and set aside.

3. Put beef brisket pieces and the remaining ginger slices into a double boiler, fill with boiling chicken broth, cover with a lid and double-boil for 1 hr.

4. Chop coriander finely. Peel turnips and cut into pieces, put into beef stock to cook for 5 mins, take out the turnips, strain, and then put into beef brisket soup to double boil until well-flavoured. Add in a little salt and pepper to taste, and then sprinkle some coriander on top. Serve hot.

豉椒涼瓜炒牛肋條粒

Stir-fried Beef Rib
with Bitter Melon and
Chili Black Beans

預備時間：**30 分鐘** ■烹調時間：**10 分鐘** ■份量：**4 人**

■ Preparation time：**30 mins**
■ Cooking time：**10 mins**
■ Serves：**4**

材料：
牛肋條5兩 (200克)
涼瓜8兩 (300克)
紅椒1/2隻
葱1條
蒜蓉1湯匙
豆豉1湯匙

醃料：
生抽1茶匙
生粉1/2湯匙
黑胡椒碎1/4茶匙

芡汁：
水3湯匙
豆瓣醬1/2茶匙
糖1/2茶匙
生粉1/4茶匙
蠔油1茶匙

Ingredients:
200g beef ribs
300g bitter melon
1/2 red chili
1 sprig spring onion
1 tbsp minced garlic
1 tbsp preserved black beans

Marinade Ingredients:
1 tsp soy sauce
1/2 tbsp cornstarch
1/4 tsp ground black pepper

Thickening Sauce:
3 tbsp water
1/2 tsp chili bean sauce
1/2 tsp sugar
1/4 tsp cornstarch
1 tsp oyster sauce

TIPS 貼士：

牛肋條可在凍肉舖購買到，多是來自巴西，注意要選多肉少筋少肥，也可當作牛腩烹煮，如切成細粒快手大火兜炒，口感有如食牛仔骨肉，既爽脆又有牛肉的甘香。

Beef ribs, mostly from Brazil, are widely available in frozen food stores. Select the meaty ones with little tendons and fat. You may cook it like beef brisket or cut it into dices for stir-frying on high heat to make a crunchy and tasty dish.

做法：

1. 牛肋條橫切1/2厘米厚片，加醃料同醃30分鐘。
2. 涼瓜開半，挖去瓜瓤，斜切件，加1/2湯匙鹽拌勻醃10分鐘，放滾水內拖水1分鐘，取出，沖凍水備用。
3. 紅椒去籽切角；葱切度；蒜頭剁蓉，與豆豉一同壓爛。
4. 燒熱鑊，下2湯匙油，加牛肋條炒透，潛酒，加1茶匙老抽炒至乾身，盛起。
5. 燒2湯匙油，先慢火爆香蒜頭、豆豉，然後把所有材料回鑊大火炒透，加芡汁炒至乾身，上碟。

Method:

1. Cut beef ribs against the grain into 1/2 cm thick slices, mix with marinade and leave for 30 mins.
2. Cut bitter melon in half, scrape its seeds and membrane, cut into pieces, mix with 1/2 tbsp of salt and then marinate for 10 mins. Blanch in boiling water for 1 min, take it out, rinse with cold water and then set aside.
3. Remove seed of red chili, cut into pieces. Cut spring onion into sections; mince garlic, and then mash with black beans.
4. Heat up a wok, add in 2 tbsp of oil, put in beef ribs to stir-fry, splash in some wine, add in 1 tsp of dark soy sauce to stir-fry until sauce condenses, and then take it out.
5. Heat up 2 tbsp of oil, stir-fry garlic and black beans on low heat, return all the ingredients to cook on high heat to stir-fry thoroughly, add in thickening sauce to stir-fry until it condenses and then transfer to a dish.

煨滷牛䐗

Stewed **Beef Shank**

預備時間：**5分鐘** ■烹調時間：**2小時** ■份量：**4人**
- Preparation time：**5 mins**
- Cooking time：**2hrs**
- Serves：**4**

材料：
急凍金錢膑2條

滷水料：
薑3片
葱2條
八角2粒
冰糖1兩（40克）
紹興酒2湯匙
生抽3/4杯
水5杯

煨滷汁：
薑2片
葱2條
滷水汁1杯
老抽2湯匙
紹酒2湯匙
糖3湯匙

Ingredients:
2 frozen beef shanks

Stewing Sauce:
3 slices ginger
2 sprigs spring onion
2 star aniseed
40g rock sugar
2 tbsp Shaoxing wine
3/4 cup soy sauce
5 cups water

Braising Sauce:
2 slices ginger
2 sprigs spring onion
1 cup stewing sauce
2 tbsp dark soy sauce
2 tbsp Shaoxing wine
3 tbsp sugar

◥**TIPS 貼士：**

滷牛膑最好選擇巴西金錢膑，因金錢膑帶有不少筋，經過炊滷後雪凍切薄片，口感爽脆，如果喜歡鬆軟口感，可多煮30分鐘，煮好趁熱食便可。

Select the beef shank with more tendons for its crunchiness. After stewing, the shank can be sliced thinly and served. If you like a softer texture, you may stew it for 30 minutes more.

做法：

1. 牛膑解凍洗淨，放半鑊滾水內拖水5分鐘，取出沖洗乾淨。

2. 預備滷水材料放煲內煮滾，加入牛膑蓋好以慢火煮90分鐘，熄火後浸片刻，保留滷水汁。

3. 預備煨滷汁材料爆香及煮滾，加入牛膑煮至汁濃稠，最後拌入麻油，盛起待涼，入冰箱雪凍，食前取出切成薄片，再澆上少許滷水汁及麻油享用。

Method:

1. Defrost beef shank, wash thoroughly, blanch in a wok half-filled with boiling water for 5 mins, take it out and then rinse thoroughly.

2. Put stewing sauce ingredients in a pot and bring it to a boil. Add in beef shank, cover with a lid and cook on low heat for 90 mins. Turn off the heat and leave it to soak for a while. Keep the stewing sauce for later use.

3. Stir-fry braising sauce in a wok and bring it to a boil. Add in beef shank to cook until the sauce condenses. Finally, put in sesame oil, transfer to a dish, leave to cool and then chill it in the refrigerator. Cut into thin slices before serving and then serve it with a little stewing sauce and sesame oil.

番茄滑蛋煮牛肉

Beef with **Tomatoes** and Egg

預備時間：**30 分鐘** ■ 烹調時間：**10 分鐘** ■ 份量：**4 人**
Preparation time：**30 mins**
Cooking time：**10 mins**
Serves：**4**

材料：
牛柳3兩（120克）
番茄3-4個
洋葱1/3個
蒜頭2粒
雞蛋1隻

醃料：
糖1/2茶匙
牛抽1茶匙
生粉1/2湯匙
水1湯匙
胡椒粉及麻油少許
油2茶匙（後下）

調味料：
水1/2杯
鹽1/2茶匙
糖1湯匙
茄汁2湯匙

芡汁：
生粉1/2茶匙
水1湯匙

Ingredients:
120g beef fillet
3-4 tomatoes
1/3 onion
2 garlics
1 egg

Marinade Ingredients:
1/2 tsp sugar
1 tsp soy sauce
1/2 tbsp cornstarch
1 tbsp water
A little pepper and sesame oil
2 tsp oil (to be added later)

Seasonings:
1/2 cup water
1/2 tsp salt
1 tbsp sugar
2 tbsp ketchup

Thickening Sauce:
1/2 tsp cornstarch
1 tbsp water

◥TIPS 貼士：

番茄去皮方法很簡單，先用尖刀切去蒂部，底部剝十字，放滾水內浸1分鐘，取出，浸水過冷再用刀輕易將皮撕去，食時口感會更好。牛柳比較腍軟，如果無加鬆肉粉的話最好將醃肉時間再加長，效果會更好。

To remove the peel of tomatoes: simply slit a cross on the bottom of tomatoes, blanch them in boiling water for 1 minute, rinse in cold water and then peel off their skin easily with a knife. To tenderize the beef, you may mix it with meat tenderizer or marinate it with seasonings for a longer period.

做法：

1. 牛肉橫紋切薄片，加醃料拌勻同醃30分鐘。

2. 洋葱切幼絲；蒜頭拍扁；番茄去皮，切成大塊。

3. 鑊內燒1湯匙油，爆香蒜頭及洋葱，加番茄及調味料蓋好煮5分鐘至軟身。

4. 雞蛋打散；牛肉加油拌勻，放番茄內煮至九成熟，再倒入蛋汁拌成蛋花，最後卜生粉水埋芡至稠身，｜碟。

Method:

1. Cut beef fillet against the grain into thin slices, mix with marinade and leave for 30 mins.

2. Shred onion; smash garlics slightly; peel tomatoes and then cut into large pieces.

3. Heat up 1 tbsp of oil, stir-fry garlic and onion, add in tomatoes and seasonings, cover with a lid to cook for 5 mins until softened.

4. Whisk eggs; mix steak with oil, put into the wok of tomatoes to cook until almost done, pour in egg liquid and stir into egg flakes. Finally, add in thickening sauce and sitr until it thickens. Transfer to a dish to serve.

沙嗲肉眼筋炒芥蘭

Stir-fried Kale and
Pork Loin Tendons
with Satay Sauce

預備時間：**15分鐘** ■烹調時間：**10分鐘** ■份量：**4人**
■ Preparation time：**15 mins**
■ Cooking time：**10 mins**
■ Serves：**4**

材料：
豬肉眼筋4兩 (150克)
芥蘭12兩 (450克)
紅蘿蔔花8片
薑6片
蒜頭1粒
沙茶醬1湯匙

醃料：
糖1/4茶匙
生抽1/2湯匙
生粉1茶匙
胡椒粉及麻油少許
油1/2湯匙 (後下)

調味：
滾水1杯
糖1湯匙
紹興酒1湯匙
雞粉1茶匙

芡汁：
水3湯匙
糖1/4茶匙
生抽1/2茶匙
生粉1/2茶匙

Ingredients:
150g pork loin tendons
450g kale
8 slices red carrot (carved into floral shapes)
6 slices ginger
1 garlic
1 tbsp satay sauce

Marinade Ingredients:
¼ tsp sugar
½ tbsp soy sauce
1 tsp cornstarch
A little pepper and sesame oil
½ tbsp oil (to be added later)

Seasonings:
1 cup boiling water
1 tbsp sugar
1 tbsp Shaoxing wine
1 tsp chicken powder

Thickening Sauce:
3 tbsp water
¼ tsp sugar
½ tsp soy sauce
½ tsp cornstarch

做法：
1. 肉眼筋橫切2厘米闊件，加醃料拌勻醃10分鐘。
2. 將2片薑切花，蒜頭切片，芥蘭修去硬梗及菜，斜切厚片。
3. 燒1湯匙油，下4片薑及芥蘭炒香，加調味料蓋好煮2分鐘，盛起，隔去水分。
4. 燒2湯匙油，下蒜片、薑花及肉眼筋爆炒至七成熟，加入沙茶醬爆香，芥蘭回鑊，潷酒，注入芡汁兜炒至入味，上碟。

Method:
1. Cut pork loin tendons into slices of 2 cm thick, mix with marinade and leave it for 10 mins.
2. Cut 2 slices of ginger into floral pattern; cut garlic into slices; trim the hard skin leaves of kale and then cut diagonally into thick slices.
3. Heat up 1 tbsp of oil, add in 4 slices of ginger and kale to stir-fry, add in seasonings, cover with a lid to cook for 2 mins, take it out and then strain.
4. Heat up 2 tbsp of oil, add in garlic slices, ginger and pork loin tendons to stir-fry until almost done. Add in satay sauce to stir-fry, return kale to wok, splash in some wine, add in thickening sauce to stir-fry until aromatic. Transfer to a dish to serve.

南瓜欖角蒸排骨

Steamed Pork Ribs with **Pumpkin and Black Olive**

預備時間：**20 分鐘** ■烹調時間：**15 分鐘** ■份量：**4 人**

■ Preparation time：**20 mins**
■ Cooking time：**15 mins**
■ Serves：**4**

材料：
腩排8兩（300克）
南瓜5兩（200克）
油欖角½兩（20克）
蒜茸1½湯匙
滾油2湯匙
葱粒1湯匙

醃料：
生粉2茶匙
鹽½茶匙
糖1茶匙
生抽1茶匙
胡椒粉、麻油少許

Ingredients:
300g pork ribs
200g pumpkin
20g black olive
1½ tbsp minced garlic
2 tbsp boiling oil
1 tbsp spring onion dices

Marinade Ingredients:
2 tsp cornstarch
½ tsp salt
1 tsp sugar
1 tsp soy sauce
A little pepper and sesame oil

TIPS 貼士：

怕排骨肥可選用肉排，但有時會韌，所以可用鬆肉粉醃過，其實用帶少許肥的腩排會較鬆軟，蒸時可加點南瓜或芋頭墊底吸收肉汁及油份，肉會不覺肥膩，南瓜又美味，是很好的配搭。

Some people like to use lean pork ribs for health reasons. However, they become tough after steaming. As a remedy, you may marinate the ribs with meat tenderizer. On the other hand, lean belly ribs are more tender in texture. Therefore, it is recommended to steam the ribs on a plate lined with pumpkins or taro to absorb the meat juice and oil secreted. In this way, you can enjoy both tender but not greasy meat and tasty pumpkins with meat juice.

做法：

1. 腩排斬細件，加醃料拌勻，同醃20分鐘。
2. 油欖角沖洗乾淨切碎；與蒜蓉加滾油拌勻，再與排骨撈勻。
3. 南瓜去皮及籽，切成長件；放碟中，排骨鋪面，隔水大火蒸12分鐘至熟透，取出，灑上葱粒，趁熱上桌。

Method:

1. Chop pork ribs into small pieces, mix with marinade and leave it for 20 mins.
2. Wash black olive thoroughly and then chop finely; mix minced garlic with boiling oil, then mix it with pork ribs thoroughly.
3. Peel pumpkin and remove the seed, cut into long pieces; line pork ribs orderly in the centre of a plate, steam above water on high heat for 12 mins until cooked, take it out, sprinkle some spring onion on top and then serve hot.

洋葱煎豬扒

Pan-fried Pork Chops
with Onions

預備時間：**20 分鐘** ■烹調時間：**15分鐘** ■份量：**4人**

■ Preparation time：**20 mins**
■ Cooking time：**15 mins**
■ Serves：**4**

材料：

有骨豬扒10兩 (400克)
洋蔥1個
蒜蓉2茶匙
生粉1湯匙 (後下)

醃料：

鹽1/3茶匙
糖1/2茶匙
生抽2茶匙
生粉1/2湯匙
水1/2湯匙
胡椒粉及麻油少許

蠔油芡汁：

水1/2杯
糖1/2茶匙
生抽1/2湯匙
蠔油1湯匙

芡汁：

生粉1茶匙
水2湯匙

Ingredients:
400g pork chops
1 onion
2 tsp minced garlic
1 tbsp cornstarch (to be added later)

Marinade Ingredients:
1/3 tsp salt
1/2 tsp sugar
2 tsp soy sauce
1/2 tbsp cornstarch
1/2 tbsp water
A little pepper and sesame oil

Thickening Sauce with Oyster Sauce:
1/2 cup water
1/2 tsp sugar
1/2 tbsp soy sauce
1 tbsp oyster sauce

Thickening Sauce:
1 tsp cornstarch
2 tbsp water

TIPS 貼士：

很多人怕豬扒不易熟透，往往會煎過火而令豬扒變乾又韌。其實最好選帶有少許肥肉的有骨豬扒，因肉質較鬆軟，還有要煎前再上少許粉，目的是幫助鎖住肉汁，先用大火，後轉中慢火煎，還可用剪刀剪開測試生熟程度。

To ensure that pork chops are fully cooked by pan-frying, many people overcook it, making the pork tough and dry. It is recommended to use the pork chops with bone and a little fat for their tender texture. Before pan-frying, coat them with a little cornstarch to block the juice inside. Pan-fry the pork on high heat and then lower the heat to medium. You may also cut it open with a pair of scissors to check if it is cooked.

做法：

1. 豬扒用刀背輕輕拍鬆，切開半，加醃料同醃20分鐘。

2. 洋蔥切條；豬扒灑上生粉拌勻。

3. 燒2湯匙油，放下豬扒以中火煎香，反轉另一面再煎至七成熟，加入洋蔥條及蒜蓉炒香，加芡汁蓋好以慢火煮3分鐘，開蓋，加芡汁兜炒至汁料香濃，便可上碟。

Method:

1. Slightly pound pork chops with the back of a knife to tenderize the meat. Cut in half, mix with marinade and leave it for 20 mins.

2. Cut onion into strips; sprinkle cornstarch on pork chops and mix thoroughly.

3. Heat up 2 tbsp of oil, add in pork chops to pan-fry on medium heat, turn it over to pan-fry until almost done. Add in onion strips and minced garlic until aromatic. Add in thickening sauce, cover it with a lid and cook on low heat for 3 mins. Remove the lid, add in thickening sauce to cook to stir-fry until it condenses. Transfer to a dish to serve.

梅菜蒸肉餅

Steamed Minced Pork
with Preserved Cabbage

預備時間：**15分鐘** ■烹調時間：**10分鐘** ■份量：**4人**
Preparation time：**15 mins**
Cooking time：**10 mins**
Serves：**4**

材料：
枚頭豬肉6兩 (240克)
甜梅菜2兩 (80克)
薑米1茶匙

醃料：
糖1/4茶匙
生抽1/2湯
生粉1/2湯匙
水2湯匙
胡椒粉、麻油少許

調味：
2湯匙油
1茶匙糖

Ingredients:
240g tenderloin pork
80g sweet preserved cabbage
1 tsp minced ginger
Marinade Ingredients:
1/4 tsp sugar
1/2 tbsp soy sauce
1/2 tbsp cornstarch
2 tbsp water
A little pepper and sesame oil
Seasonings:
2 tbsp oil
1 tsp sugar

做法：

1. 豬肉切條後切粒，然後剁碎，加醃料順一方向攪成膠狀。

2. 梅菜洗淨浸水5分鐘，搾乾水，切碎，加薑米及調味料拌勻隔水先蒸5分鐘至糖溶化，待涼後加豬肉內再攪至稠身。

3. 將肉餅鋪平在碟上，隔水蒸10分鐘至熟透，待片刻，取出，便成一道佐膳家常小菜。

Method:

1. Cut pork into strips and then dices. Mince the pork, mix with marinade by stirring unilaterally into a glue.

2. Wash sweet preserved cabbage, soak for 5 mins, pat dry and then cut into fine dices. Stir in minced ginger and seaonings and mix well. Then steam above water for 5 mins until sugar melts. Leave to cool and then add in pork to mix until sticky.

3. Place meat cake on a plate, steam above water for 10 mins until cooked, leave for a while, take it out and a homemade dish is done.

菠蘿咕嚕肉

Sweet and Sour Pork
Belly with **Pineapple**

預備時間：**25 分鐘** ■烹調時間：**20 分鐘** ■份量：**4人**
- Preparation time：**25 mins**
- Cooking time：**20 mins**
- Serves：**4**

材料：

肥腩排6兩 (240克)

青椒1/3隻

紅甜椒1/3隻

洋蔥1/4個

菠蘿2片

蒜頭1粒

蛋黃1隻

生粉1/2杯

甜酸汁：

水150毫升、糖1/2磚片

茄汁2湯匙、白醋3湯匙

鹽1/8茶匙、生粉1茶匙

水1湯匙

醃料：

鹽1/4茶匙

糖1/4茶匙

生抽1茶匙

酒1茶匙

生粉1湯匙

Ingredients:

240g fatty spare ribs

1/3 green pepper

1/3 red pepper

1/4 onion

2 slices pineapple

1 garlic

1 egg yolk

1/2 cup cornstarch

Sweet and Sour Sauce:

150ml water

1/2 cube sugar

2 tbsp ketchup

3 tbsp white vinegar

1/8 tsp salt

1 tsp cornstarch

1 tbsp water

Marinade Ingredients:

1/4 tsp salt

1/4 tsp sugar

1 tsp soy sauce

1 tsp wine

1 tbsp cornstarch

◥TIPS 貼士：

傳統的咕嚕肉是用肥豬肉做，所以我把腩排修切出來的肥腩肉，經長時間炸透後變得更鬆化脆口，不覺肥膩。咕嚕肉的甜酸汁要夠大甜大酸，煮時要用長時間煮到夠濃稠，近乎糖膠狀才加入少許生粉水埋芡，汁料仍不宜太多，否則會令咕嚕肉變腍。

Traditionally, this dish is made of fatty pork. Therefore, I have used the fatty spare ribs for its fat. After thorough deep-frying, the fatty meat will become crispy and spongy. Meanwhile, to make a strong sweet and sour sauce, you have to cook it long enough until it reduces into a syrup. You may then add in some cornstarch liquid to thicken. However, don't make too much sauce or it will soften the crispy pork pieces.

做法：

1. 腩排修切成細件，加醃料拌勻同醃20分鐘。

2. 蒜頭切片；洋蔥、青、紅椒及菠蘿切角。

3. 腩排加蛋汁拌勻，續粒沾上生粉，用手握緊，放滾油內炸至熟透，盛起待片刻，回滾油翻炸至金黃色，盛起。

4. 燒1湯匙油，爆香蒜片、洋蔥、青、紅椒，盛起。

5. 鑊內加甜酸汁以中火推煮至汁呈半糖膠狀，蔬菜回鑊煮熱，埋少許芡汁，炸腩排回鑊快手兜勻，上碟，趁熱品嚐。

Method:

1. Cut fatty spare ribs into small pieces, mix with marinade and leave for 20 mins.

2. Slice garlic; Cut onion, green and red peppers, and pineapples into triangles.

3. Mix spare ribs with egg yolk, coat with cornstarch, squeeze tightly with hands, put into a wok of boiling oil to deep-fry until cooked, take it out and leave for a while. Return the pork to the boiling oil until browned, and then take it out.

4. Heat up a tbsp of oil, stir-fry garlic slices, oninon, and green and red peppers, and then take it out.

5. Put sauce ingredients to a wok to cook on medium heat until it turns into a syrup. Return the vegetables to the wok, pour in thickening sauce, return deep-fried pork to wok to stir-fry, and then transfer to a dish to serve.

照燒豬軟骨撈米線

Teriyaki Pork Cartilage
with Rice Noodles

預備時間：**15 分鐘** ■烹調時間：**90 分鐘** ■份量：**4 人**

■ Preparation time：**15 mins**

■ Cooking time：**90 mins**

■ Serves：**4**

材料：

急凍豬軟骨1¼斤 (750克)
蒜蓉3湯匙
乾葱蓉4湯匙
洋葱粒1杯
照燒醃醬4湯匙
茄汁6湯匙
紹興酒1湯匙
雞粉1茶匙
滾水2杯
米線1包

醃料：

老抽2湯匙
紹興酒1湯匙
胡椒粉少許

Ingredients:

750g chilled pork cartilage
3 tbsp minced garlic
4 tbsp minced shallot
1 onion
4 tbsp teriyaki marinade
6 tbsp ketchup
1 tbsp Shaoxing wine
1 tsp chicken powder
2 cups boiling water
1 packet rice noodles

Marinade Ingredients:

2 tbsp dark soy sauce
1 tbsp Shaoxing wine
A pinch of pepper

TIPS 貼士：

豬軟骨是近胸關節部份修切出來帶有白軟骨的肉條，經過炆煮後，連骨也可吃掉，廣為認識的是豬軟骨拉麵。豬軟骨可在大型凍肉公司買，惟貨源不多，主要多交食肆用，所以我會挑選金沙骨較肥軟的骨條代替，這個肉較多，炆煮時間較長，但效果很相近，最好炆煮後再焗過夜，那麼便有肉鬆骨軟效果。

Pork cartilage is the white dense connective tissue cut out from the rib cage. It becomes so soft after stewing that even the bones are edible. It is available at large frozen food stores and is commonly used for making soup noodles. However, as most of it is sold to restaurants, it is short in supply. Therefore, I will use the fatty and soft spare ribs as a replacement. Although the ribs take longer to cook, they contain more meat and taste like cartilage. The ribs should be stewed and then left in a covered pot overnight to tenderize before serving with noodles.

做法：

1. 豬軟骨洗淨，放滾水內拖水5分鐘，盛起，趁熱加醃料同醃。

2. 用鍋燒熱2湯匙油，下乾葱、蒜蓉及洋葱爆香，加豬軟骨、照燒醃醬及茄汁爆透，注入酒、雞粉及滾水煮滾，以慢火炆80分鐘至腍身及汁濃稠。

3. 燒大鍋滾水，下米線煮熟，沖凍水過冷河，再用滾水沖過，瀝乾水，上碟，加豬軟骨伴吃，便成惹味豬軟骨。

Method:

1. Wash soft pork cartilage thoroughly, blanch in boiling water for 5 mins, take it out and then mix with marinade.

2. Heat up 2 tbsp of oil, add in shallot, minced garlic and onion to stir-fry. Add in pork ribs, teriyaki marinade and ketchup to stir-fry thoroughly. Pour in wine, chicken powder and boiling water, and bring it to a boil. Then simmer on low heat for 80 mins until the meat is tender and the sauce condenses.

3. Heat up a large wok of water and bring it to a boil. Put in rice noodles to cook, rinse with cold water and then rinse with hot water, strain and then transfer to a dish. Serve with pork ribs.

蜜桃沙拉骨

Pork Ribs with
Honey Peach

預備時間：**10分鐘**（另醃排骨1小時）■烹調時間：**90分鐘**
■份量：**4人**
■ Preparation time：**10 mins** (plus 1 hr for pork rib marination)
■ Cooking time：**90 mins**
■ Serves：**4**

材料：
一字排10兩 (400克)
水蜜桃2件
生菜2片

醃料：
乾葱蓉2茶匙
鹽1/4茶匙
糖1/2茶匙
美極醬油1湯匙
喼汁1½湯匙
生粉1湯匙
蛋汁1湯匙 (後下)
2湯匙生粉 (後下)

沙律醬汁：
奇妙醬4湯匙
茄汁1茶匙

Ingredients:
400g pork ribs
2 pcs honey peach
2 slices lettuce

Marinade Ingredients:
2 tsp minced shallot
1/4 tsp salt
1/2 tsp sugar
1 tbsp Maggi seasoning sauce
1/2 tbsp Worcestershire sauce
1 tbsp cornstarch
1 tbsp beaten egg (to be added later)
2 tbsp cornstarch (to be added later)

Salad Dressing:
4 tbsp miracle whip
1 tsp ketchup

TIPS 貼士:

沙拉骨有點西餐風味，所以要用美極醬油及喼汁同醃才合襯，汁烹排骨要用多油炸才易熟，回鑊加醬汁同兜炒時不可開大火，否則奇妙醬會還原，變成油份。

Pork ribs have to be marinated with Maggi sauce and Worcestershire sauce to bring out the flavour of this fusion dish. Remember to deep-fry the ribs in a large amount of oil to get it cooked thoroughly. When the ribs are stir-fried with the sauce, remember to cook it on low heat or Miracle Whip will return to an oily state.

做法：

1. 一字排修切成2厘米×4厘米長件，吸乾水，加醃料及乾葱蓉同醃1小時。

2. 水蜜桃切厚片，預備醬汁料備用。

3. 排骨炸前拌入蛋汁及生粉，放滾油內炸至金黃色及至熟透，瀝乾油份。

4. 鑊內燒少許油，排骨回鑊兜炒至熱透，熄火，加入醬汁及蜜桃片兜勻，放生菜面，即成一夏日醒胃小菜。

Method:

1. Cut pork rib into strips of 2cm x 4cm, pat dry, mix with marinade ingredients and minced shallot, and leave it for 1 hr.

2. Cut honey peach into thick slices; prepare salad dressing and set aside.

3. Mix pork ribs with beaten egg and then cornstarch, put into hot oil to deep-fry until browned and cooked, and then strain.

4. Heat up a little oil in a wok, return pork ribs to stir-fry until hot, and then turn off the heat. Add in salad dressing and honey peach to stir-fry, and then transfer to the top of lettuce. A refreshing dish for the summer is ready to serve.

繡球肉丸

Pork and **Shrimp**
Meatballs

預備時間：**35 分鐘** ■烹調時間：**15 分鐘** ■份量：**4 人**

■ Preparation time：**35 mins**

■ Cooking time：**15 mins**

■ Serves：**4**

材料：

免治豬肉3兩 (120克)
蝦肉3兩 (120克)
馬蹄粒 2湯匙
金華火腿蓉1湯匙
蛋2隻
小棠菜4棵

醃料：

鹽¼茶匙
糖⅛茶匙
生粉½湯匙
水½湯匙
胡椒粉、麻油少許

芡汁：

清雞湯⅓杯
生粉½茶匙

Ingredients:
120g minced pork
120g shrimp meat
2 tbsp water chestnuts
1 tbsp minced Jinhua ham
2 eggs
4 crops Shanghai white cabbage

Marinade Ingredients:
¼ tsp salt
⅛ tsp sugar
½ tbsp cornstarch
½ tbsp water
A little pepper and sesame oil

Thickening Sauce:
⅓ cup chicken broth
½ tsp cornstarch

◣ **TIPS 貼士:**

煎薄蛋皮切忌多油，所以先要加約4湯匙油落鑊燒熱，把梣個鑊｜滿油，再把多餘油全部倒走，調校成中火，蛋汁倒入鑊中央，把鑊搖晃，上滿鑊成 大片薄蛋皮，煎至離鑊便可取出切絲。

Don't use too much oil to pan-fry the egg sheet. Just heat up 4 tbsp of oil in a pan, spread the oil evenly on the pan and then pour out excess oil. Turn the heat to medium, pour egg liquid at the centre of the pan, shake the pan slightly to spread out the egg liquid into a large circular form, fry until the egg sheet is detached from the pan, take out the egg sheet and then shred.

做法：

1. 蝦肉挑去黑腸，洗淨，吸乾水，拍成蝦蓉。
2. 豬肉加醃料攪至起膠，再加火腿蓉、馬蹄粒及蝦蓉攪至起膠，分成十份，搓成小丸子，備用。
3. 雞蛋拂勻，加少許鹽調味，用少許油煎成薄蛋皮，取出，切成幼絲。
4. 把每粒丸子滾上蛋絲，置抹油碟上隔水蒸12分鐘，把丸子移放碟上。
5. 燒半鑊滾水，加少許鹽、糖、油將菜拖熟，排繡球肉丸旁。
6. 燒少許油，下芡汁煮滾，澆肉丸上即成。

Method:

1. Remove intestines of shrimps, wash thoroughly, pat dry and then mash into shrimp paste.
2. Mix pork with marinade and stir until sticky, add in minced ham, water chestnut dices and shrimp paste and stir in a glue. Divide into 10 portions, knead into small balls and set aside.
3. Beat eggs, add in a little salt, and then pan-fry into a thin egg sheet with a little oil. Take it out and then cut into shreds.
4. Roll the meat balls over shredded fried eggs, transfer to a plate brushed with oil to steam above water for 12 mins. Transfer the meat balls to another plate.
5. Boil half wok of water, add in a little salt, sugar and oil, put in vegetables to blanch until cooked and then line them around the meat balls.
6. Heat up a little oil, add in thickening sauce and bring it to a boil. Pour it over the meat balls and serve.

雜菌燴豬柳

Braised Pork Tenderloin with Assorted Mushrooms

預備時間：**15分鐘** ■烹調時間：**15分鐘** ■份量：**4人**
■ Preparation time：**15 mins**
■ Cooking time：**15 mins**
■ Serves：**4**

材料：
豬柳6兩（240克）
鮮冬菇3兩（120克）
鮮蘑菇3兩（120克）
雞髀菇4兩（150克）
三色椒3兩（120克）
乾葱1粒
花奶5湯匙
牛油1湯匙

醃料：
鹽1/4茶匙
糖1/4茶匙
生抽1茶匙
生粉4茶匙
胡椒粉、麻油少許
蒜茸1茶匙

芡汁：
水1/2杯
雞粉1茶匙
鹽及糖各1/8茶匙
生粉1/2湯匙
胡椒粉少許

Ingredients:
240g pork tenderloin
120g fresh shiitake mushrooms
120 fresh button mushrooms
150g fresh abalone mushrooms
120g shredded red, green and
yellow peppers
1 shallot
5 tbsp evaporated milk
1 tbsp butter
Marinade Ingredients:
1/4 tsp salt
1/4 tsp sugar
1 tsp soy sauce
4 tsp cornstarch
A little pepper and sesame oil
1 tsp minced garlic
Thickening Sauce:
1/2 cup water
1 tsp chicken powder
1/8 tsp salt; 1/8 tsp sugar
1/2 tbsp cornstarch
A pinch of pepper

TIPS 貼士:

豬柳是豬肉最瘦又腍軟部份，最適宜老幼食用，切絲肉皆較鬆散，切件前宜加芡汁煮，清淡得像食牛仔肉般，效果很不錯。

Pork tenderloin is the leanest and most tender part of pork, and thus, is most suitable for all. If you cut it into shreds, it will loosen up the meat and ruin its texture. By pan frying it and then cooking it with a white sauce, the pork tastes like a light-flavoured veal.

做法：

1. 豬柳橫紋切1厘米厚片，加醃料及蒜茸同醃。

2. 鮮冬菇、蘑菇開半；雞髀菇切片；青、紅、黃三色椒切條；乾葱切片。

3. 燒滾水，放下鮮菇拖水1分鐘，盛起，瀝乾水。

4. 燒3湯匙油，下豬柳煎至九成熟，加乾葱片及三色椒爆香，潷酒，三菇回鑊加芡汁煮滾，最後拌入花奶及牛油煮至汁濃稠，可跟白飯或意粉同吃。

Method:

1. Cut pork tenderloin against the grain into slices of 1 cm thick, mix with marinade and minced garlic.

2. Cut fresh shiitake mushrooms and button mushrooms in half; cut abalone mushrooms into slices; cut green, yellow and red peppers into strips, cut shallot into slices.

3. Boil some water, put in fresh shiitake mushrooms to blanch for 1 min, take it out and then strain.

4. Heat up 3 tbsp of oil, pan-fry pork tenderloin in it until almost done, add in shallot slices and pepper strips to stir-fry, splash in wine, add in the mushrooms and thickening sauce to stir-fry until it boils. Finally, add in evaporated milk and butter until the sauce thickens. Serve it with rice or spaghetti.

（一）調味料介紹
Seasonings

泰國酸辣雞醬
Sweet Chili Sauce for Chicken

OK汁
OK Sauce

美極醬油
Maggi Sauce

鰹魚粉
Bonito Powder

玫瑰露酒
Rose Essence Wine

橄欖菜
Preserved Cabbage in Oilve

照燒醃醬
Teriyaki Marinade

日本青芥末
Wasabi

沙茶醬
Satay Sauce

（二）料頭的切法
Cutting of Condiments

葱度
Spring Onion Sticks

薑絲
Shredded Ginger

薑蓉
Minced Ginger

薑花
Ginger in Floral Shape

葱絲
Shredded Spring Onion

青、紅椒角
Peppers in Triangle Shape

葱粒
Chopped Spring Onion

洋葱絲
Shredded Onion

蒜片
Sliced Garlic

蒜蓉
Minced Garlic

洋葱粒
Onion Dices

芫荽片
Chopped Coriander

辣椒仔
Red Chilis

乾葱碎
Chopped Shallot

乾葱片
Sliced Shallot

（三）常備料頭預先加工
Preparation of Condiments

對於一些家庭主婦來説，有時候如可預先製成部份料頭，到入廚時可隨手可得，方便快捷，如：

It is a good practice amongst households to prepare some condiments beforehand so that they are readily available for use, for example:

薑汁酒
Ginger Wine

可用半斤去皮薑切粒，放入攪拌機內加一杯雙蒸白酒磨碎，隔渣，便成薑汁酒，入樽。放冰箱內可保存一星期。

用途：薑汁酒適用於醃雞、醃豬膶及任何內臟、醃魷魚及炒芥蘭等。

Peel 300g of ginger, cut into dices, grind it with a cup of double distilled wine in a food blender, and then filter out the residues. The ginger wine thus collected is then bottled and stored in the refrigerator for a maximum of 1 week.

Usage: Ginger wine can be used to marinate chicken, pig liver, internal organs of poultry and livestock, squid and stir-fry kale.

油浸蒜蓉
Minced Garlic in Oil

蒜頭去衣，沖淨，吸乾水，放小型碎磨機內磨碎；燒熱油，注入蒜茸
內，拌勻，入樽。放冰箱內可保存一星期。

用途：蒜蓉適用於炒菜、炒肉、醃豬扒、牛扒、雞扒。

Skin the garlic, wash, pat dry, and then grind with an electric food grinder.
Heat up some oil, pour it into minced garlic, stir thoroughly, bottle and then
store in the refrigerator for a maximum of 1 week.

Usage: Minced garlic can be used for stir-frying vegetables and meat, as
well as marinating pork chops, steak and chicken fillet.

葱粒
Chopped Spring Onion

葱去頭去外衣，沖淨，吹乾，可切度，切絲或切粒，放保鮮盒內可保存
三至四天。

用途：適用於蒸魚、炒肉、煮粉麵或灑小菜面裝飾。

Remove the bulb and skin of spring onion, wash thoroughly, blow dry, and
then cut into sticks, shreds or dices. It can be kept in a sealed container and
stored for 3 or 4 days.

Usage: As accompaniment for steamed fish, stir-fried meat and noodles or
garnishing.

（四）煮食方法的運用
Cooking Methods

煎
Pan-Frying

用少量油把食材煎熟，令食物的表面甘香及熟透。煎的火候可用大火、中火或細火，視乎食材質地，例如豆腐便要用大火，因豆腐含水份多，所以要用高溫令水份盡快揮發，才可把豆腐煎至金黃香口。另外煎肉應先以大火，當肉汁鎖定後便可轉中弱火煎熟，煎糕點或翻熱食物便用小火，以免把食物燒焦。

Use a small amount of oil to pan-fry food until browned and cooked. Adjust the level of heat according to the texture of food, for example, tofu has to be cooked on high heat. Due to high water content of tofu, it has to be pan-fried on high heat to let its water evaporate quickly and make its surface brown and crispy. On the other hand, meat has to be pan-fried on high heat first to block its juice from seeping, and then continue to fry on low heat until the meat is cooked. For the reheating of Chinese pudding or meat, low heat has to be used or the food will be burnt easily.

炒
Stir-frying

以少量油用高溫快手把食材兜炒至熟透。炒肉要避免黏鑊。首先，開大火將鑊燒至有微煙，加入適量油，把油上滿鑊，下肉在鑊中央，待煎片刻，肉有微甘香後便反轉另一面略煎，然後才快手兜亂，這樣可把肉汁鎖住，肉食便會外香內嫩，這便是中菜的精髓。中式炒菜也要講求鑊氣，同樣火力要猛，下菜後最好要加蓋蓋片刻，把菜的水份迫出來，這樣炒出來的菜能保留原味，口感又夠爽脆。不過，炒餸菜通常要分兩個階段完成才會完美，第一階段是將食材炒至八成熟透，第二階段是再起鑊爆香料頭如薑、蒜，葱等再把食材回鑊加汁炒透，依照這個做法在家中也可做到"小炒王"的鑊氣，日日有好餸。

Stir-fry the food ingredients with small amount of oil quickly on high heat until cooked. To avoid food sticking, you have to: heat up the wok until smoke comes out, add in some oil, shake the wok slightly to get it coated with oil fully. Put in the meat at the centre, leave it to pan-fry briefly until the downward side is slightly browned, flip it over to pan-fry the other side, and then stir-fry the meat quickly to block the meat juice. In this way, the meat surface is browned while its meat remains succulent.

On the other hand, vegetables also have to be stir-fried on high heat. When vegetables are put into the pan, we have to cover it with a lid to let it cook briefly so as to induce the secretion of water from vegetables. This can preserve the taste, texture and crunchiness of vegetables. However, stir-frying dishes with vegetables has to be carried out in two stages: firstly, stir-fry the food ingredients until almost cooked; secondly, stir-fry the aromatic ingredients, such as, ginger, garlic and spring onion thoroughly and then return the food ingredients to the wok and add in sauce to stir-fry until cooked. In this way, you can make delicious dishes of restaurant quality at home.

炸
Deep-frying

以大量食油炸熟食物。要視乎食物而決定油溫和炸食物的時間，通常要先將油溫提升至沸點，可用木筷子來測試，若有氣泡沿筷子升起便是沸騰了，也可用其中一件食物來測試。當所有食物加入炸油後應要調較至中火，用足夠時間炸熟食物，過大火炸會使食物外觀很快變得金黃但內裏卻未熟透。另一個容易出現的問題是油膩，緊記當食物炸至九成熟後可轉用大火將食物內裏的油迫出來，即時撈起，不應等熄火後撈起，否則食物會即時吸收油份，令食物變得油膩了。剩下的油可待涼透及沉澱後隔乾淨，留作煮食時再翻用。

Deep-frying is to cook food in large amount of oil. The temperature of oil and the cooking period have to be adjusted according to the food ingredients. Normally, the oil must first be heated to its boiling point. To ensure it is hot enough, test it with a wooden chopstick. If bubbles emerge along the chopstick, it means the oil is boiling. You may also put in a piece of food to test the temperature of the oil. When all the food has been put into the oil, the heat has to be lowered to medium to allow enough time for the food to cook. Otherwise, the food will be browned quickly while its inside remains raw. In addition, when the food is almost done, remember to cook on high heat in order to force the oil out of the food, and then take out the food immediately when the heat is still on. If not, the oil will be instantly absorbed by the food once the heat is turned off, making the food very greasy. The remaining oil can be cooled and filtered for future cooking.

炆
Stewing

以較多的汁料，用中細火和較長時間烹煮食材至腍軟。一般多用於質地較韌或體積較厚大的材料，例如牛腩、羊肉或鴨肉。有時也可以用燉的方法去代替，時間要再加長，燉的食材會較原汁原味，炆煮餸菜會汁濃肉鬆，各有好處。

Stewing uses lots of sauce to cook the food on medium to low heat for a longer period of time until the food is cooked and tender. This method is usually used for cooking food ingredients of a tough texture or large in size, for example, beef brisket, mutton or duck. It can sometimes be replaced with double-boiling, which takes a longer cooking period. Double-boiling can preserve the original taste of the food while stewing can make the meat tender in a strong sauce.

灼
Poaching

用上湯或滾水，在短時間內以大火煮熟材料，能保持食物的鮮味和爽脆嫩滑的口感。例如白灼蝦、灼清菜等。

Poaching is a quick method to cook food with broth or boiling water on high heat. It can maintain the fresh taste and delicate texture of food and is suitable for making dishes of simple fresh ingredients, such as, poached shrimps and vegetables in broth.

煲
Hard-Boiling

將材料放入大量滾水中，經大火煲滾後再轉用中細火用長時間把材料煲至出味和腍軟，例如煲湯、煲粥等。

Put the ingredients into large amount of boiling water and cook on high heat. When it is boiling, lower the heat to low or medium to simmer for a long period time until the taste of food comes out and the food becomes tender. This is normally used for making soup and congee.

蒸
Steaming

用大火煮滾水，利用蒸氣的熱力把食物蒸熟。首先，待水滾後，把食物放蒸架上，蓋好，隔着水蒸熟，例如蒸魚，但蒸蛋卻忌用大火，否則蒸蛋便成蜂巢狀的老蛋。

Steaming is to cook food with the heat generated from boiling water. Firstly, bring the water to a boil, place the food on the steaming rack, and then cover it with a lid to steam above water. This method is generally used for steaming fish. For cooking eggs, never steam it on high heat or the egg will be overcooked.

新手入廚學中菜 Beginner for Chinese Cooking

編著	Author
鄭慧芳	Lilian Cheng

編輯	Editor
郭麗眉	Cecilia Kwok

翻譯	Translator
葉翠顏	Tracy Ip

攝影	Photographer
幸浩生	Johnny Han

設計	Designer
任霜兒　馮麗珍	Annie F　Adrianne Feng

出版者　Publisher
萬里機構‧飲食天地出版社　Food Paradise Publishing Co., an imprint of Wan Li Book Co. Ltd.
香港鰂魚涌英皇道1065號東達中心1305室　Room 1305, Eastern Centre, 1065 King's Road, Quarry Bay, Hong Kong.
電話　Tel: 2564 7511
傳真　Fax: 2565 5539
網址　Web Site: http://www.wanlibk.com

發行者　Distributor
香港聯合書刊物流有限公司　SUP Publishing Logistics (HK) Ltd.
香港新界大埔汀麗路36號中華商務印刷大廈3字樓　3/F, C & C Building, 36 Ting Lai Road, Tai Po, N.T., Hong Kong.
電話　Tel: 2150 2100
傳真　Fax: 2407 3062
電郵　E-mail: info@suplogistics.com.hk

承印者　Printer
美雅印刷製本有限公司　Elegance Printing & Book Binding Co. Ltd.

出版日期　Publishing Date
二〇一三年四月第八次印刷　Eighth Print in April 2013

ISBN 978-962-14-3858-4

萬里機構wanlibk.com